The New Earth Reader

Editorial Board

Editorial Address

Terra Nova
New Jersey Institute of Technology
Newark, NJ 07102
(973) 642-4673
FAX (973) 642-4689
e-mail: terranova@njit.edu

The New Earth Reader

The Best of *Terra Nova*

Edited by David Rothenberg and Marta Ulvaeus

The MIT Press
Cambridge, Massachusetts
London, England

This book was set in Berkeley oldstyle book by Wellington Graphics.

Printed and bound in the United States of America.

Library of Congress Cataloging-in-Publication Data

The new earth reader : the best of Terra Nova / edited by David Rothenberg and Marta Ulvaeus.

    p. cm.

  Includes bibliographical references.

  ISBN 0-262-18195-9 (hc : alk. paper)

   1. Environmentalism. I. Rothenberg, David, 1962–   .

 II. Ulvaeus, Marta. III. Terra Nova (Cambridge, Mass.)

GE195.N48 1999

363.7—dc21   99-34983

                                                                CIP

Contents

Acknowledgments

Terra Nova would never have come into being without the encouragement of Madeline Sunley, who at the time was environmental studies editor at MIT Press. Sabine Hrechdakian, our managing editor for the first two years, worked carefully on many of the pieces included here and shepherded them into print. The New Jersey Institute of Technology has provided a warm and supportive home for us since inception, and we'd especially like to thank Norbert Elliot, chair of the Humanities and Social Sciences Department, and John Poate, dean of the College of Arts and Sciences, for believing in us from the start.

We'd also like to thank all the members of our editorial board and in particular Sean O'Grady, who served as poetry editor, staff astrologer, and confidante over the years.

Student assistants have included Mia Söderlund, Cecilia Kelnhofer-Feeley, Inna Ososkov, and Cara Mia Ciasulli.

And thanks to all of our readers, from our loyal subscribers to those who prefer to thumb through copies at the local megastore. Thanks to the *Utne Reader, Harper's,* and the Council of Educational and Learned Journals for excerpting from our pages and for honoring us.

Algimantas Kezys

Prologue: The *Terra Nova* Dream

I dreamt I was naked on the ice. It was cold, and the snow was burning against me. I didn't know where I was. Pages were flying all around. Pages marked up with red. The proofs. I was correcting the proofs of *Terra Nova*, volume 1, number 1.

"You can't call your magazine *Terra Nova*," said a cool, deep voice. It was Robert Falcon Scott. We were lost in the Antarctic. "That's the name of my ship. Get up, get some clothes on. We've got to keep moving. We've just left the South Pole and we're heading home."

"Wait a second," I thought. I knew we weren't going to make it. We were destined to freeze to death out here. This expedition was the famous example of bad judgment that had resulted in failure. We had reached the Pole, but no one would survive.

I did as I was told, pulling on some warm sealskins. But I was there for a reason: to protest. "I want the name," I said. "Eighty-five years from now I will start a magazine. It will be called *Terra Nova*."

"Can't do that," said the Captain. "If this expedition makes it, the name will refer to the glory of our mission, our boat, forever. If we don't . . ." [he was beginning to see the future] "then it will be a term of disgrace, insidious failure."

"But the new world, the new land, the new earth? How can the hope inside those words ever be lost? The future will need hope even more than the past." I tried to make a case.

"So you know about the future?" He took some more interest, shaking the snow off his parka. "What will the Earth look like then? What will Antarctica become?"

Wars will come, more terrible than any imaginable. More cruelty, more extremes of destruction. The destruction of the Earth will be called an "environmental crisis," and we will all be held accountable. Antarctica will still seem largely untouched, but greedy countries will carve it up, penguins will be burnt as fuel, trash will be strewn about the Pole. It will be the last wilderness, as the world wears itself out . . . I told him what I thought was enough.

"And you want to start a *magazine* in the midst of all this? What use are words at the end of the world?"

This was the question.

The ice went black. It was time to get up, but I stayed in the dream so I could claw out an answer to him:

> I don't have the answer, but won't abandon the questions.
> Time is running out, but we don't know how to stop it.
> There are too many views on each problem—
> Subtlety will not be dissolved into polemics.
> We must not despair, we must evoke, and celebrate.
> Celebrate the poetry in the mundane.
> Recognize the nature in all things,
> not just the wilderness, not just the green,
> but also the soot, the gray, the black of smoke and the yellow of poison.
> Human nature: a connection to matter that makes human life possible—
> progressing, transforming, rebuilding the world,
> yet never forgetting the rhythms and desires we have been given. No one
>   is completely free nor completely determined.
> That's just the way it is.

"So what does your magazine have to do with nature?" he asked, picking the ice out of his eyebrows.

"Everything, and nothing," I said. "The whole distinction is a human idea. Like Snyder says: 'No nature.' Saying no, and at the same time wanting to know. 'We do not easily *know* nature, or even know ourselves . . .' The greatest respect we can pay to nature is not to trap it, but to acknowledge that it eludes us . . ."

"All right, but you're eluding me. Now let's get up. We have to march north."

And I followed the leader's order. I buttoned up my snowsuit, preparing to walk into history and face what could only be a certain death.

Terra Nova. It's that New World, the place of hope and possibility on the other side of the ocean, the indistinct future waiting for human transformation. It's Scott's ship on the ill-fated Antarctic expedition. It's a name no one can own. It comes from an ancient language, yet it speaks plainly of the unknown future. What is new? what is old? How can one be sure of these things?

Everything that connects humanity to the world around us brings the mind closer to nature. We begin as part of nature, but as we continue, we see and describe it as separate from us, finding new ways to both fit in and run away from where and what we are. We are neither necessary to the Earth as witnesses, nor are we a blight that will destroy the planet. But we are here nonetheless, and we look to nature for example, instruction, salvation, and for food and fuel.

We have had to narrow the idea of nature in order to turn it into something fragile that we need to protect. But nature is also powerful, greater than our knowledge or use of it can ever be, formidably *beyond* our possible conceptions of it. There is more to the world than any of us can ever know. And that belief is necessary for knowledge to grow and push forward, and for evolution to continue.

Environmentalism has always been a curious mix of the old and the new. A romantic notion of past and distant cultures as being closer to the Earth juxtaposed with visions of an enlightened tomorrow when the whole world will share values of respect for our habitat and join together to make sure we will not run it down.

Who can love a word like "environment," so technical and far from the real immediacy of our surrounding home? "Nature" is hardly better, with its all-encompassing ambiguity. There are choices that seem more tangible: ocean, air, life, sky, green, blue, cloud, wind, ground. Yet the intricacies are still one step removed from the sudden call. Environmentalism, ecology, conservation—these words have become politicized. But the affinity for nature is not a single point on the left-right political scale; it is a deeper cultural tendency that will not go away.

My intention when I started *Terra Nova* was not to go after the sensational, the factual, or even particularly the relevant immediacies of possible impending environmental doom or revelation. I believe in a more subtle approach. It is my feeling that the connections between humanity and nature are far more diverse, mysterious, and confusing than most ecological writing has been willing to admit. I don't believe in easy answers. Nor do I believe in the truisms of nature writing, in which polemicists travel all over the world and urge everyone else to stay home.

Enough admonishment. There is ambiguity, magic, struggle, effort, and confusion in making peace with the more-than-human world. Nature is the strongest challenge we have. It is never easy to face it as it is: indifferent, resilient, impervious to word or image. I want to see new ways of celebrating and criticizing nature, ones that have not been given voice before. I want readers to be surprised and challenged, to be thrown together from different camps and asked to reckon with conflicting sides and irreconcilable views. No environmental problem is simple, no encounter between the wild and the tame can be seen in just one way.

Old stories will become new when told in new ways. New stories will work if we remember them forever. Neither optimism nor pessimism is appropriate. What is called for is an honest attempt to push toward the *truth,* recognizing how hard it is to be honest in the placing of words upon experience. The fact of nature is at the heart of what it means to be human. Nature abhors its own absence. It will always be there. There are few human creations in which nature doesn't have a place. Technology, medicine, design, mathematics, psychology, and religion—each one of these challenges us human beings to take stock of its context, what surrounds it. What is us, and what is not? It is hard to be sure.

In describing this venture, I have made much of my desire to cross borders with the work published here. The blending of different genres—say, environmental thought and literature—brushes against so many standards of credibility. Each side expects something else, and the walk between them is a fine line—to ensure that the work will be appreciated and not be allowed to fall between the cracks.

There is also the border between academia and the rest of the world, perhaps imagined, and sometimes widened by those standing on opposite sides. It is certainly true that the academy has always encouraged among its members a kind of hermetic style of writing, which only those within its borders are meant to understand. Strong intellectual work should be made comprehensible to the nonspecialist. Some pieces of writing may at first be difficult and take some effort to read, but they should be worth the effort. It has certainly been my objective to toss different kinds of writing together, to show that there are many ways of expressing awareness of the mystery that lies in the confluence of the civil and the wild.

I am not in favor of popularizing stories, nor of sensationalizing stories to make them more appealing. The world is interesting enough as it is, if we know how to pay attention. This journal has brought together different kinds of work: fiction, analysis, poetry, reportage, essays, and art on all aspects of the human-natural continuum.

We can't escape environmental problems except by solving them. And after each small step forward there is still more to discover. We can celebrate the unknown if we admit it is there.

Of course, you don't have to agree with any of this to fit in. You're already here.

Welcome to the best of *Terra Nova*.

—David Rothenberg, editor

D. B. Cooper, Where Are You Now?

John P. O'Grady

> *D. B. Cooper, where are you now?*
> *We're looking for you high and low.*
> *With your pleasant smile*
> *And your dropout style,*
> *D. B. Cooper, where did you go?*

In the Pacific standard midafternoon of November 24, 1971, Northwest Airlines Flight 305 was about to embark in rain from the Portland of Oregon, to Seattle, Washington, last leg of a long and prosaic milk run begun that morning in the nation's capital. Outside, this close to the ground, the weather was a factor. Taking the middle seat in the last row on the right side of the plane was a "nondescript middle-aged man" in a business suit. He carried a briefcase, and he gave his name as Dan Cooper, afterward misidentified by the press as "D. B. Cooper"—testimony perhaps to his overall slipperiness. Even his alias was elusive.

   After settling in like any other passenger, he placed his briefcase on the unoccupied aisle seat beside him. Soon after takeoff, he handed the stewardess a folded piece of high-quality bond paper. Unceremoniously, she stuffed it into her pocket. "I thought he was trying to hustle me," she would later report. He shook his head, gestured for her to read it. He moved his briefcase to the unoccupied window seat beside him. Digging deep into her pocket past sudden dread, she retrieved the folded piece of high-quality bond paper. She unfolded it. It was a typed note in plain prose: "Miss—I have a bomb here, and I would like you to sit by me."

She sat by this nondescript middle-aged man in a business suit. She looked at him. He was wearing dark glasses. He had a briefcase. She looked out the window, saw sky above the clouds; up here, the weather was not a factor. The man opened his briefcase on the unoccupied window seat beside him. Next the stewardess was shown his bomb: a large battery connected by a maze of copper wire to eight long cylinders of what looked like dynamite. He closed the briefcase. He spoke: "Would you please return that note to me?" His voice was gentle; he was polite. She handed to him the wavering note typed on high-quality bond paper. He slipped it, unwavering, into the inside pocket of his suit jacket. It was a fine weave of wool. What now? Soft as clouds he said, "Take dictation, a note for the pilot." What choice but to turn scribe, embrace complicity? She had to provide her own paper, but she was about to become coauthor of what must rank among history's best-paid pieces of writing. Words flowed from his mouth. Words flowed through her hand. Shaky penmanship would suggest an uneasy collaboration: "I want $200,000 by 5:00 P.M. In cash, in twenty-dollar bills. I want two back parachutes and two front parachutes. Make them sport parachutes. When we land, I want a fuel truck ready to re-fuel. No funny stuff, or I'll do the job." She had no interest in funny stuff. She did her job, delivered the note to the pilot. When she returned to her seat, the nondescript middle-aged man said thank you. Later, this note in the steward-ess's hand would prove to be the only surviving text of D. B. Cooper.

What followed in the gathering gloom of that grim and chill November eve-ning was circling, three hours of it, over Sea-Tac Airport; it took time for Northwest Airlines to come up with the two hundred thousand dollars—in stacks of twenties—but not time enough for the FBI to come up with an effec-tive response to this man wearing a suit. All agreed: This was a novelty in the annals of skyjacking. Law enforcement was at a loss. They had been drawn be-yond the pale. Nobody had a map. Here they were without protocol. Before this, no hijacker had ever asked for a parachute. Before this, no hijacker had ever demonstrated such precision, such cordiality, such grace in executing his escape. Before this, no hijacker had ever commandeered a plane for so pure a motive as D. B. Cooper's. He was not commonplace. He was not violent, rude, or slovenly. He had no complex political motives. He simply wanted the money. Bold, bright, and clean—this was the clarity of greed! Something everyone could understand.

Among the workaday, his was a technique that would be envied, even publicly admired, which is why he became a folk hero. In after years, a Portland librarian was known to confess, "I've thought about ways to get a lot of money like he did, but I would never really do it, so I'm glad for him." Not uncommon either was the sentiment of a betweeded professor of literature who said, for the record, "Anyone who has the guts to parachute out of a jet in the middle of a dark and stormy night has my admiration. I hope he got away with the money, and I hope he's not dead."

Apparently, none or few of the passengers aboard Flight 305 on that dark and stormy night knew what was going on. The pilot informed them that the plane had developed mechanical problems, that they were circling "in order to burn up excess fuel before landing"—ordinarily most disconcerting news, but apparently not to these fated passengers, who were remarkably content in their circling, in their calm burning of "excess fuel" in the unsettled night sky over Seattle, Washington. It seems that even without complimentary drinks, they were set at ease by their captain's voice; something in its immanent authority they found soothing. Everything was under control. They could sit back and enjoy their circling. Reported one of the passengers later, "He could make you feel good even if you were walking toward a guillotine."

At last the plane came back to earth, touched down at Sea-Tac in early evening and steady rain. The money and parachutes were delivered, the thirty-five passengers and two of the three stewardesses were released. D. B. Cooper could not afford release of the pilot and copilot: Somebody had to fly the plane. Nor could he afford release of the stewardess: Somebody had to be the hostage. When his inspection of the chutes revealed standard military issue, D. B. Cooper became noticeably peeved. He had ordered sport parachutes, the sort that enabled extended free falls from high in the sky. These workhouse military chutes would not do. With an artist's rancor, he sent them back, along with the name of a skydiving shop where the proper type of chute could be obtained. All of this delayed things. Beyond the tarmac, however, officials who did not necessarily like each other nodded gravely. In retrospect, all agreed that, except for the parachutes, the events were well scripted.

While everybody waited for the proper chutes to arrive, the nondescript middle-aged man in a business suit kept his middle seat in the last row on the right side of the plane. His briefcase remained on the unoccupied window seat beside him. He instructed the remaining stewardess to sit in the aisle seat beside him. The money was in a heavy canvas bag stenciled with the words SEATTLE FIRST NATIONAL; it rested nearby in the aisle itself. D. B. Cooper had taken off his dark glasses; his eyes were said to be blue. The stewardess looked out the window, now dark; she could see drops of rain streaking down on the outside of the glass. D. B. Cooper moved his hand to the inside pocket of his suit jacket, removed another folded piece of high-quality bond paper, and gave it to the stewardess. In his soft cloud voice he said, "Please take this forward to the flight crew, have them read it, then return it to me."

She cooperated, delivered the dispatch to its intended audience. It was a typed note with no mistakes. It directed the crew to fly the plane to Mexico City. "Fine," the pilot said in his unwavering voice, "but tell that guy this plane can't go that far in one jump. We'll need to make two interim fuel stops." The pilot had no typewriter—his was an oral transmission. The stewardess hastened back with the pilot's message. D. B. Cooper listened politely to her news. She then returned the typed note. "Fine," said the unruffled man, now folding the used note with care, now returning it to the inside pocket of his suit jacket, now withdrawing another folded piece of paper. This time it wasn't high-quality bond. "Please deliver this to the pilot," he said. It was a standard-issue government form, typed and meticulously filled out: a federal aviation flight plan. "Then return it to me," he added.

When this latest curiosity reached the forward cabin, the flight crew puzzled over its precision: *"Follow flight path Victor 23 south, toward Reno, Nevada."* Quizzical looks were exchanged. D. B. Cooper's route purposely avoided the wild woolly wags of the Cascade mountains. Temperatures there would be well below freezing; snow would be falling. In the lowlands, there would be rain. What kind of hijacker was this, both experienced in the diving of sky and possessed of extraordinary skills in navigation?

At one point, a rain-lashed official in a trench coat walked up to the door of the plane. He was from the Federal Aviation Administration, but he was very wet. He requested permission to board, apparently in an attempt to reason with the skyjacker. The hard things are glorious. Permission refused. D. B. Cooper now suggested to the pilot, via another specimen of that now disturbing high-quality bond paper: "Let's get this show on the road. Return this note to me." It was a moving communication. The proper chutes suddenly arrived. The note was returned. The show once again took to the air.

Onward to Reno. Not long into the dark sky, D. B. Cooper's hand returned to the inside pocket of his suit jacket. The stewardess shuddered. D. B. Cooper withdrew another of his high-quality works: an appendix to the well-composed flight plan. This one read, "Maintain an altitude of 10,000 feet, lower the flaps to fifteen degrees, hold the airspeed to 170 knots. Return this note to me." The stewardess brought it to the flight cabin. More quizzical looks. The pilot frowned. These operating conditions were just about the minimum for keeping a 727 aloft, and at 10,000 feet in this part of the country, there were mountains out there to worry about. But D. B. Cooper knew what he was doing. The flight path he had chosen kept them over the lowlands, over the open fields scored by Interstate 5.

Bewilderment now moved aft with the stewardess. She returned this last note to D. B. Cooper; he folded it and returned it to the inside pocket of his suit jacket. Then he reasonably requested her to help him open the exit door leading to the aft cabin stairwell. No typed letter, no high-quality bond paper. They were past all written words. Now it was strict performance, direct action. D. B. Cooper picked up his briefcase from the unoccupied window seat beside him. He had the stewardess lead the short walk to the aft of the cabin. The 727 is the only commercial jetliner that has a door beneath the tail. Ordinarily, it is not a good idea to open it in flight, but given the drastically reduced speed and altitude, a reasonably safe parachute jump might be made. Besides, this man claimed to have a bomb, spoke words that seemed like clouds with mountains hiding in them. Even without a typed letter on high-quality bond paper, the stewardess complied. She showed him how to open the door. Then he ushered her forward to the cockpit and locked her in with the flight crew. That was the last anybody saw of D. B. Cooper.

A little after 8 P.M., perhaps somewhere over Lake Merwin on the dammed Lewis River in southwestern Washington, a light flashed on the instrument panel in the cockpit of Northwest Flight 305, indicating that the rear exit door had indeed been opened. Dropped the cabin pressure to that of the summit of Mount Hood. Dropped the temperature to seven below. Dropped, too, it would seem, D. B. Cooper, suited for business, with two parachutes, a briefcase, and a heavy green burden. Unseen into the wild dark.

Of the three or five military jets bird-dogging the truant 727, none sighted any chute openings. The weather was inclement; sullen clouds continued to roll in from the Pacific; at 10,000 feet there was no visibility. When, on the far side of California's Sierra Nevada, Flight 305 punched through the clouds and landed at the Reno airport, D. B. Cooper was nowhere to be found. He must have jumped. A four-state manhunt was launched. All points bulletins were flashed from Nevada to Washington, nearly the entire length of the Cascade Range, but efforts were concentrated in 150 square miles of thickly resistant forest near the Washington towns of Longview and Ariel, just north of Portland. Here, authorities reckoned, was where he jumped. Here, authorities discovered, the forest was uncooperatively dense, the terrain amply crenulate, sufficient not only to hinder an effective search but to present a serious hazard to the cavalier sky jumper who descended into this thicket bereft of visibility, bereft of warmth. D. B. Cooper, on the threshold of his leap, was not known to have had ample protection from the elements. A business suit, a briefcase, a heavy canvas bag—anything could have happened. Said one member of the posse after a week of unsuccessful searching, "We're either looking for a parachute or a hole in the ground."

More than two decades have passed. No parachute, no hole. But ballads have been composed, films have been made, books have been written, imposters have come forward. The pilot and the copilot of Northwest Flight 305 have since retired. The stewardess long ago quit her job and became a nun. D. B. Cooper is still being sought.

His remains the only unsolved skyjacking in U.S. history. The weather may have been a factor.

Matthijs Maris, *Young Woman*, 1863

From *A Philosophy of Clean*

D. L. Pughe

You will laugh when I tell you that all I think about is cleaning, since you witness so little of it in the scraps of life we share. You may choose to see it as theoretical, but it is a lot closer to the floor than that.

Though I cannot imagine someone cleaning for me, I have no qualms about cleaning for others. It provides a comfortable anonymity, and because the goal is to erase the traces of disorder from daily life, it becomes the art of invisibility itself. I have cleaned for many years in many places and have come to know that *clean* is defined everywhere quite differently.

My work is confined to the day, and the evening returns me to my attic of books and drawings to celebrate the freedom I imagine I have from a wealthier life. This freedom also includes carefully conserved layers of dust and a few spiderwebs which I respectfully do not wipe away, when I do approach tampering with the surfaces of my own home.

In my attic rooms I have everything that gives me pleasure close at hand. My window box is full of colorful flowers nestled in ivy, which trails down the eaves. In the other window a birdhouse attracts a crowd of birds throughout the day. Just before dawn each morning I spread crusts on the ledge and watch the birds arrive to enjoy a noisy breakfast in a spray of seeds. They are all the small scrappy brown and gray birds of the city, with an occasional bright-chested finch unaware of its difference. Their struggle for the crusts is frantic and greedy, especially in winter, but when they have eaten their fill, they retreat to the low branches of the tree and puff out the feathers around their eyes. They sink back into fluffy balls with half-shut lids and doze for several moments, then fly away.

On Sundays, if the weather is poor, I sometimes sit by this window and write or draw and watch the birds on the other side of the glass. Once I set my music box on the sill and a salmon-breasted finch quickly arrived. It had no interest in the crumbs and stood with its head cocked looking inside, contemplating the tinny version of a Beethoven sonata. For a moment I watched its eyes, deep and still; it clearly understood something I hoped to know but was too foolishly trapped in the human world to fathom. But it also could have been pausing to see its own reflection in the glass. And how odd to be thrust from a world of invisibility into one of appearance.

Like the birds, I can choose to appear or not; I have been able to remain invisible. It is a great source of strength, though not of power.

Many of those who have employed me often spoke in my presence as though I were not there. For the couple T., their conversations were deliberations over social events, character assertions and rejoining assassinations about each of their acquaintances, though even the most maligned still appeared at their dinner engagements.

I know how much time this took them, these acrobatics of social grace. I knew by their expression when it preoccupied them and, moments later, a chance remark would reveal their captive state. I always felt immense relief that my own time is more free for the thoughts that help me through the procession of hours in each day. Ideas which I can change and exchange with what I discover, smudges of propriety I can simply wash away.

In the presence of this couple, who were both engaged in the intellectual circle of the area, I usually converted my occasional smile into a gestureless gaze, but I found my mouth turning up one evening, just as I was clearing away their coffee cups before I departed. The man of the house looked up from his newspaper while his wife chatted on about an upcoming dinner.

His glance to me was sharp, then curious.

"What amuses you?" he asked.

Startled out of my invisibility, any confession which revealed that I could hear them when they speak would be the most dangerous thing I could say. It is what makes a foreigner infinitely preferable as a servant, affording those being served a greater privacy. I had, in turn, broken my own rule about not troubling anyone with my expression.

My other rules connect to truth, but to answer his question honestly would have been to admit that private thoughts keep me company while, separate from them, I went about my work. It would have pressed me into three dimensions, too large for their image of me.

"It is this," I said, without lying, holding up the matchbox illustration of a red bird with outspread wings. "This picture always makes me smile."

The man stared at me blankly, and he looked at the box. It did not make him smile and forced him to recognize for a moment that I observe certain things, which could perhaps mean I am able to see him or his wife.

I watched this fact pass before him, followed by a mild reflection that I was someone, usually so emotionless in demeanor, for whom the world did contain ever so small delights. I saw his eyes shift from the box of matches to my face with a kind of pity, then turn back to his newspaper.

His wife looked up from her list to her husband then to me, and I glanced at the bird on the box. Its small round eyes were confident, its wings were full and strong, and I smiled again.

I first cleaned for my grandmother. She was a petite woman of great warmth and humor who had grown up in the country and carried aspects of it with her to life in the town. Nature never failed to astonish her. She knew the names and habits of every bird, she gave every living thing which grew or wandered indoors a sanctuary in her ancient plaster house. Covered in vines and with floors sagging from age, her home offered extraordinary warmth and comfort. God or nature were intertwined in her view, she instinctively nourished my desire to know and understand the world. Observation was the first path, but imagination, she insisted, can take you farther still. Books were a way to travel, languages suggested entry to realms never dreamed of.

When I was quite small, she helped me create a gypsy shelter in the overgrown garden within the high-walled courtyard behind her home. Stacked firewood formed one wall, with natural ledges and shelves. Thick honeysuckle bushes closed the back, and an old blue canvas sail propped by sticks became my roof and a third wall. My grandmother would bring me a tin pan to cook my scrounged provisions, and we would sit on a log there. She would take on a different voice, explaining an imagined vagabond life. She continually brought me books to whet my appetite for faraway places, even though this meant that I would leave her. And she left a lantern in the window of her attic bedroom so that, on nights when I insisted on trying to sleep out under the dark sky, trembling but determined to not give in to fear, I would be watched over by her surrogate eye.

I remember her skin and her touch. Toward the end, I massaged her back at night. In the beginning, she had always been affectionate with me—I arranged to sit near her chair, and she would stroke my neck and hair for hours as she played cards with the many sisters she had coaxed to move near her. They passed and dealt and sipped sherry through the furry indoor light of many afternoons. My job was to refill their tiny glasses from a crystal decanter they had labeled "The Fountain of Youth," then to disappear into the dark of the pantry to refill the decanter from a covered vat. "The decanter must never be empty," my grandmother explained, for it was filled with nectar from the legendary source which was keeping all of them young. But when I asked for a drink myself, "Not yet, not yet," the ladies clucked and laughed and then tipped their cups.

Each year as my grandmother grew weaker, I began to do more than pour. Pouring stopped, in fact. She asked me to clean instead. I found a century of cobwebs under the feet of her bathtub. The oven had a crust from years of bubbled casseroles. And the ancient tile floor of her kitchen had a tapestry pattern of stain which slowly unraveled as I washed it away.

My grandmother's most treasured porcelain plates had paintings of birds; my favorite was a luminous blue pheasant who inhabited many of my dreams. When I began to take care of her home, I realized that my grandmother had diligently washed the front of these fine dishes for years, but never the back. There was a film, almost like trampled soot, which took me a day to scrape away.

She appreciated everything I ever did for her. She was always kind. But the pale colors I revealed under the layers of dirt pleased her more than anything. More than the countless drawings and portraits I had given her which filled her halls, more than the cards I made for her, the stories I wrote. More than the nectar from the Fountain of Youth. And more than winning at cards, which she did so easily.

She taught me by her gratitude that I could make a difference through my efforts. This reassured a deep overwhelming dread that my life might not count.

I learned what it meant to have a room "smell" clean and began to explore the supplies which could produce the best smells. I experimented with waxes and finally found one which could make her floor shine like the most rare marble under a sheet of thin ice. But I would lie awake at night, worrying she might slip and all this cleanliness would be her downfall.

She fell, it turned out later, on the slick shiny floors of the hospital when she was visiting an ailing sister. It led to a long painful decline.

It is the smells that always take me back to her.

I remember raking her leaves into piles and burning them near my gypsy camp, the pungent smoke filling the air, the scent also from her hearth when I cleared out the ashes. The smell of soap as I washed the plates with birds, the scouring powder she had me use as cleanser on the bottom of her sink, in the bathtub, on the most difficult places on the floor.

The smells of clean in my grandmother's house are particular, mingling with her own scent of powder, of peppermint hard candies, of pungent sherry. Yet they are also portable. The soap, I have found, is available here in this foreign land, far from those distant mountains.

My walls are lined with books in worn paper covers lovingly arranged and re-arranged in the middle of the night. They are my treasures, along with senti-mental objects on the top of each shelf: paintings and drawings from various friends, dusty glass tubes and beakers from a laboratory, a smoked-glass box full of rolled ribbons, a pewter cup engraved with pictures from the past, a small wooden Russian church I built when I was a child, and a brass candle holder with angels that circle and dance above the heat of the flames. Each angel is engraved with a line from a philosophic fable written by my blind friend J., who brought me this treasure. He attached a crystal prism to the skirt of each angel so that when they pass over the brass bell the ring is sharp and clear, and the reflected prismatic light dances over the beams of my room. He could not see this, yet he made it for me. He imagined it so vividly it car-ried out his wishes. And he has suggested that one can envision life, then live with clarity of anticipation.

You see, I live in the greatest luxury. In the close space I need little heat and am seldom cold. I curl in my bed nestled in one of the eaves, covered by the feather comforters from my grandmother. The colored spines of my books smile down on me in the dark, everywhere I look lies evidence of friends and their thoughtfulness and talents, the capacity for human beings to open a window on meaning, one they allow us to see through.

I am able to take down each book or object at any time of the night, and this remains for me the greatest freedom, to have enough light to read by, paper to write on, and the words of these authors from all ages and all over the world, able to converse at an hour when everyone in the city is usually sleeping.

This is the deepest privacy, a delicious secret I share with myself alone.

Seeing as how I hold in most of my thoughts during the day, I seek out friends at night, always curious where our conversation will travel.

When the weather is fair, I walk.

One of my favorite places is a shallow lake now mostly overgrown with trees and shrubs. Yet today when I walked there it was full of winter resonance, an arbor of pearl gray branches with only a flag here and there of bright yellow leaf still hanging on. A wooden slat bridge sits low over the shallow water and zigzags across the lake, keeping you in suspense above and unable to see what lies ahead. In several places, an alcove is formed and there is a bench where one can conduct a private conversation. It is here I have sometimes stopped with a friend to talk.

In spring when the leaves are full, this wooden pathway becomes a cave of green, and small ducks swim in between the branches and reeds. We are held then, by wood above water, in the arms of trees. When my friend speaks to me I think of the letter Oldenburg once wrote to Spinoza, thanking him for his recent "gift of mind." It is this sensation of deepest thanks I feel when a friend shares their thoughts with me. I cannot explain it other than as a kind of honor. You will laugh knowing that when a dog chooses to come toward me, wagging, I feel honored by the recognition. A thought imparted has a different resonance, though I would not place it on a scale. To be given an idea one has not encountered, pressed into your hands in the course of conversation, is a breathless challenge leading to whole orchards more. When I return to cleaning, it is the thoughts imparted to me that I carry, anxious to begin thinking them through.

In the remote area where I was raised, up in the mountains, I realized how far ideas must travel from their source. Since most influential ideas emerge in the density of the human world, they must, like a river forced upstream, climb up to where other, simpler notions glisten like ice. And there is always the question of whether they will melt what they find, dissolving into a new way of thinking.

Up there, the world of events was often so far away, and news of it traveled to us slowly. We lived in an envelope of nature and accustomed ourselves to its time and sense. My mother and father were deeply respectful of books, of all things in print; my grandfather in a distant city was a printer. They imparted a faith that the world could be yours if you learn its secrets, and those could be just pages away.

But from isolation I came to realize that much of what I was able to learn was partial, always incomplete. It led to ideas which were only approximate and echoed our whole way of life.

Because the commercial goods of the cities were not available, much energy was put into fashioning approximations of what was too distant or too dear. My mother and father believed the world was always within reach through effort: when you discovered something you needed or wanted, you simply made it. You fathomed how to construct it, you saved for or found the materials, you made a facsimile to the best of your ability. In this way, the world was never denied us, though there was a blurry line about what constituted an original or authentic world.

And just as ideas pass from theories into words and follow into the actions of the day, a similar blurring occurs, blunting the original thought. I have often noticed the difference between who plants the seeds of ideas and who nurtures them, the way they are raised, disciplined, and cherished in the course of time. The way they disperse in the world is as telling as how they began.

In my own desire to understand, I have had to work harder to retrace the steps from the source, to arrive at the original idea. It has been a long road down out of the mountains into regions where distinctions are sharply drawn. I have learned that I should not settle for what is approximate in thought but should question the casual assumptions I might have overlooked in the past. But I also know the innocence in replicas based on truth; they are not to mislead but to be able to have a place in the world, flattering the source from which they came.

As I clean, I notice what I am gathering up, aware of what I am laying down in offerings of order; knowing when I have been partial or only approximate in my own definition of *effort* and of *clean,* wrestling with my often-discomforting need to be encyclopedic, to consider *everything* and check it more than once, and questioning my shallow satisfaction at things deceptively clear. At the same time, in another chamber of my mind, I want to be able to hold and stroke and consider a new idea, sometimes finding it echoed in the patterns on the floor.

I once cleaned for a young German woman, Frau H., who was born with a deformed hip and walked with great difficulty. She had gone to the University and studied literature and philosophy and was now married and had a young son. She would sometimes be at home while I cleaned and would move her chair in order to speak with me while I worked.

With her I learned the absolutes of clean, the deepest recesses one can comprehend of dirt and soot and residue, and also the qualities of purification, the soap and cleaners and how each is applied. She bought rare bleaches and scouring brushes I had never seen and directed me from the chair as I crawled across the floor. The fumes were penetrating, and we laughed at the way our eyes welled with tears.

"You can smell how clean it is!" she said triumphantly, and I knew that no matter what I might do in shifting the dirt on the floor, the smell would convince us it wasn't there.

But it also nearly killed me.

On the tram on the way home one evening, I couldn't breathe. I kept grasping for air. My eyes watered uncontrollably, and I staggered off the tram as though drunk, reeking antiseptically of bleach.

When I finally reached my rooms, I fell on the bed in a complete panic, unable to think, terrified that I had destroyed much of my brain.

I began to test myself, to remember passages of Spinoza's *Ethics,* the sequence of Pascal's wager, and I fell asleep in a fog of images that led to even more frightening dreams.

It was several days before Spinoza returned, or the reasons why I should be interested in him. It was a complete amnesia for those days, my mind fully emptied out but without the peace of a cleansing absence. Everything had been blown away as if by a tornado, with a similar sense of loss.

Frau H. did not force me to use these disinfectants at the same strength a second time, and we slowly found a balance of what was necessary to actually clean and what was required for her scented definition of freedom from fear of the unclean.

The dirt that can be seen is often the least of our worries. The great unseen—the minute molecules of decay and disorder—linger in crevices and wait for the first chance pooling up of liquid to dive in and multiply in a wild bacchanal.

The unseen subscribes to no exponential formula, reproducing as fast as the imagination will allow.

My own hands, and those of my friends who use theirs for work, have a life of their own. They have an instinctive knowledge of how clutter defines the realms of the soul and how to set it right.

Sometimes it troubles me that cleaning is seen as the least valuable activity to be paid for in time, but I've been lucky. I have cleaned mostly for women who were responsible for too much already, women who were physically unable to look after that part of their home. They appreciated my efforts and shared parts of their lives with me in a way that I trust.

Frau H., who desired an antiseptic cloud surrounding her home, was also one of the most rare scholars. She sat in her chair while I worked and shared her theories in literature and philosophy. She breathlessly told me about Büchner's *Lenz* and read the opening pages in such an emotional voice that we couldn't continue:

> On the 20th of January Lenz went across the mountains. The summits and the high slopes covered with snow, grey stones all the way down to the valleys, green plains, rocks and pine trees. It was damp and cold; water trickled down the rocks and gushed over the path. The branches of the pine trees drooped heavily in the moist air. Grey clouds travelled in the sky, but all was so dense—and then the mist rose like steam, slow and clammy, climbed through the shrubs, so lazy, so awkward. Indifferently he moved on; the way did not matter to him, up or down. He felt no tiredness, only sometimes it struck him as unpleasant that he could not walk on his head.

Lenz, in his first mystical experience on the mountainside, also gave me the first clues I could piece together about my own private experience in nature as something neither quaint nor restful but powerful and extraordinary and unutterable.

As a child, the words I troubled over at night were ordinary words, words like *the same* and *always*. *The same* suggested a repetition which was exact and would become monotonous and was as frightening a punishment as I could imagine. When combined with *always*, it was horrifying, doing the same thing, over and over into a vague endless Eternity I could not even contemplate. For I realized that the things which gave me pleasure were never quite the same. But chores or tasks shared the greatest resemblance in their repetitive nature; this was why they became duties instead of desires. I had yet to learn I could control how I performed each act and that nothing is ever truly the same.

At first, though, the continuous acts of repetition in cleaning were deeply disturbing; they haunted me and lowered me from wonderment into the everyday. I was forever filled with longing for a "last time" I would ever have to clean a sink or stoop over a hearth. I was anxious for time and all of its detritus to stop in an unimpeded glistening silence. But repetition, Kierkegaard reminded me, "is like an indestructible garment that fits closely and tenderly, neither binds nor sags." Gradually I began to see the differences in what I was faced with and then found how many ways I could approach doing things with distinct variations. My fears began to soften, then disappear. Kierkegaard insisted that modern life held a more genuine repetition than the past. The dilemmas we face make the fact of repeating what is repeatable into the act of making something new. Repetition, he believed, can allow us to become conscious of our actual life at the exact moment we are involved in doing something we have done so many times before.

Ordinary "newness" I have come to see as a distraction, that sparkling unblemished nature of products which offer a shiny veneer for our magpie eyes. I find myself most comfortable in clothes and plates and books which have been handled countless times before, respectfully welcoming the enduring presence of the people in their past. And when the world does seem leaden, I am always comforted, again and again, by the look and sound of a stream of water as it rattles into a pail.

Now that I am living away from the village in the mountains, I am able to steep myself in anonymity. I have gradually given away all of the carefully handmade clothes from my past, facsimiles created from what was available, based on fashion pictures which made their way up the steep incline. For all their earnest creation, they were off, their *faux* nature blared and drew attention. I began to wear used clothing from secondhand shops, indulging in their scents, taking delight in and wondering over the tags of manufacturers sewn inside. I carefully selected everything I wore in order to disappear into this new realm, thankful after years of the open scrutiny of a small community that I could find privacy in the street.

In this transition, I tried to understand what it is in attire which sets people apart. I notice that inexpensive clothes are tight or too baggy, made in rigid sizes, where more expensive ones fit from being custom-made. Cheaper garments usually have too much attached to compensate—a bit too much lace, one too many colors, a cut which is attention-getting but in an unflattering line. Then the textures of cloth: those with the finest weave, the softest feel, retain shape, whereas the cheaper fabrics have wide uneven threads which sag or tear, and the surface begins to erode quickly into bumps or snags.

Because of the number of things I have washed and pressed and folded, I find it difficult to hold up a new garment without imagining it transformed after each washing into a different form. Because all things become misshapen in time, I have found which clothes hold up, the ones that bear repetition. When I go in search of my own clothing, I look for cast-off garments which I know will last. And because I am now in this anonymous city, instead of the small unprivate town, I know I might pass those who may have worn them before. But I have also grown safe in their silence.

At the end of the week, I sleep in a little later and spend time reading near the birds. Then I dress carefully and walk through the narrow streets of the neighborhood, through the stairs that wind between houses which hover over one another, until I reach the hotel, a gigantic white chateau of towers and glass.

I pick a small table near a window as close as possible to the mountain that rises up behind it, but also modestly set in a corner where I may observe the interior easily. I like one close enough to the panes of glass to be able to immediately forget the bustling interior when one of the large birds, a raven perhaps, sails down from a tower in a slow arc and lands on the grassy lawn before the pond. He conducts himself there as though he were the owner, inspecting the upkeep of a property which he had assigned to a second party but in which he still maintains a cautionary interest.

I order tea and wait with great anticipation for it to arrive, for this is the most formal moment of indulgence I have in a week, a secret compact with an elegance I can afford but which won't detain me. And, waiting, I look around the hotel, the visitors lounging in the velvet chairs, the dark wood and thick carpets, the cushioned silence. The apparent calm masks the trembling restlessness of this place, where thousands of people are continually passing through.

I have not cleaned room after room in a hotel, but when I am there I always imagine it. I am aware of the residue left behind, the ways in which people treat objects, sheets, the bath, and the floor when they know that someone will soon follow to restore it. I picture gathering up the crumpled towels, the occasional toiletries forgotten on the counter.

I think one could begin to put together a story: the hair in one bath one day surfacing in the bath of a nearby room, commingled with that of its usual inhabitant. The clues of sheets. Every kick and turn, passionate rising and falling, could finally be deciphered in their folds. Or the scents which one would draw away from knowing.

In cleaning, it is best not to know too much, wanting only the suggestion of what may have taken place.

It is in these fleeting moments of leisure that I always find myself drawn back into the web of my working life, unable for a moment to forget what my hands travel over by habit, or the nature of my place in the world. I cannot walk by a large home without imagining cleaning it, I am not able to think of going up and ringing the bell without a servant's expression on my face, nor can I avoid exchanging a knowing glance when another servant opens the door and notices my coat.

I have spent time trying to build up my claims, to reassure myself that this life, the one I now know how to control, will not get the better of me. That all my ambitions, which aim toward understanding rather than acclaim, will fall flat from my very servility. I crave freedom but not power and have been unable to fathom how to succeed as the world defines it, within my conscience and not attempting to rise above anyone else. I have chosen instead to work and comfort my hope with a vital though only immediate knowledge. On certain exhausted nights, this hope flickers wildly, threatening to go out.

When I pass by the housekeeping carts in the hotel, I am often tempted to stop and ask the women there,

"How do you get through the flood of rooms each day? What are you able to accomplish at night?"

I want to find out if they, too, are sometimes afraid.

And if they are not, how have they loosened this fear?

When I was just nineteen, I saved all my money for months from work as a server in a small café and rented a cabin in a forest on the edge of a lake high in the mountains. I had decided to write a great philosophical treatise on the moral decay and economic imbalances of our time. I was using Descartes's *Meditations* as my model, a type of mental cleaning by cutting away every fraction of doubt. I spent days wandering through the forest, formulating these deep thoughts, and pausing to write in my little book. Soon a hummingbird who lived in the forest became alarmed at my frequent appearances. These birds have firm territory; they know the enemy only by the rosy color of alarm. I wore a red kerchief on my head as I walked, and the hummingbird took to soaring over my head, swooping down just grazing my scarf with enough beak to make me take shelter in the trees.

I had to find an escape through the dense brush of the forest, dodging his attacks, to return to the empty cabin with its quilted coverlet and tin pans. I had not felt loneliness such as this before. My gypsy fantasies had failed to include it. Growing up so close to nature, I had never imagined myself apart from it, unwelcome, and able to cause alarm—as though the natural world achieved its own order and cleanliness only when people were absent.

Gradually, I stopped writing and was only good at listening and then only heard the hollow rush of wind through the trees.

I had no way to rejoin nature then; I was an intruder at best. At worst, I was human, a fact I cannot change and have never forgotten. Where Descartes came to take comfort in human "superiority" in his quest, I found the opposite. I had come not to take apart my mind, but to take apart society, and this continually led to the odd materialism and self-serving behavior which colors humanity. And the hierarchies which evolve, where one person's time is thought to be more valuable than another's, end up going beyond social distinctions into the worth of the person as a whole.

I could not extract myself from this web and stain, realizing that I, in turn, have valued other creatures less than myself. I became the damn spot that won't come out. The whole world tilted at this point and became a tribunal in which I could not excuse myself from the chain of despair. I stopped eating because I felt it was no longer right to take away the fiber of any living being, plant or animal. And I waited for a sign that I was meant to exist, yet without any faith in what might appear.

After a few days, I resigned to leave and let my scarf go in a gust of wind in the meadow, watching it head into the darkened forest and disappear. I took the train away from the mountains, walking the final distance to the house of my grandmother, hoping she would take me in and offer me something warm to eat. She was sitting on her sleeping porch in a rocking chair, smiling. Over our meal of toast and cheese, I suddenly forgot my loneliness on the mountain and renewed the theoretical quest I had abandoned. I asked her for advice on some Latin terms for my treatise, how to say with acerbic intent,

"I own, therefore I am."

She smiled with all of her silver teeth,

"Habeo ergo sum."

I wrote quickly, then drifted off, lost and exhausted.

As she rose to go into the house, my grandmother turned and smiled. I remember a wink.

"Amo ergo sum."

That was her answer.

My foolish earnestness, I realized, could become a constant hazard. I had failed to question thoroughly the "nature" of my intentions in writing; the direction of my doubt was as dubious as my premise. What I was truly hoping to understand was far from what I was experiencing, thinking, feeling in the course of each day. Her reply comes back to me now as this reminder.

I often have a dream within a dream—one where I see myself wake and rise in my long white nightgown and wander through the darkness from the small bedroom under the eaves into the long narrow parlor of the attic lined with books and its two windows tucked under the roof where the birds come to eat. As I drift in, the light is blue, and there is a glow from the silver bowl of decorated eggs painted for me by my friend D., the most unusual patterns and designs. The bowl, sitting on the table near the window ledge, holds the eggs, which are quivering. The muted moonlight creeping through the broad leaves of the tree falls across the table, embracing the bowl.

I wake and rise in my long white nightgown and walk through the darkness from my bed under the eaves to the narrow parlor, to the window where moonlight falls across the silver bowl. The eggs are broken, and there, lining the windowsills, are the most extraordinary small birds, whose feathers bear the colorful markings of their decorated shells.

I wake and rise and, in my long white nightgown, walk through the darkness from my bed under the eaves. I rise to confirm the rumor of magnificent freshly hatched birds in the narrow parlor, on the windowsills, where moonlight falls across their decorated shells filling the silver bowl.

And there they are, lined up, twitching, a bit clumsy, peeping a little, waiting to be let out into the early morning chill.

I wake and rise and, in my long white nightgown, walk through the darkness from my bed under the eaves. I reach the windowsills in the moonlight of the narrow parlor where I slide the glass open and watch the fluttering tentative flight of all the moist, brightly patterned feathers as they rise beyond the broad leaves of the tree, above the fading blue shadows, above the cool pungent scent of morning which rushes upward, above the silence ringing in the silver bowl of quivering shells.

Will the Real Chief Seattle Please Speak Up?:
An Interview with Ted Perry

David Rothenberg

> *Every part of the earth is sacred. . . . All things are interconnected. What happens to the earth happens to the sons and daughters of the earth. . . . Man did not weave the web of life, he is merely a strand in it. Whatever he does to the web, he does to himself. . . . Where is the thicket? Gone. Where is the eagle? Gone. What is it to say goodbye to the swift pony and the hunt?* The end of living and the beginning of survival.

These chilling fragments come from the famous speech of Chief Seattle, probably the single best known summation of the ideas of the environmental movement. It is familiar all across the globe, appearing in countless talks, epigrams, and quote books. It is said to be compulsory for German schoolchildren to memorize the thing, and in Scandinavia there are Chief Seattle Clubs dedicated to its message. At the museum beneath Mount Rushmore, there once was (and perhaps still is) an interactive diorama with the disembodied head of a Native American that glowed red when you pressed the button, and the solemn words came out in a soothing, serious voice, impossible to forget.

The words of Seattle are profound, inspiring, and the stuff of a truly spiritual document. They touch people in the way religious texts are meant to, deep inside the heart, straight down to the feet and the hallowed ground beneath—wherever you are. For the words teach us that all ground can be seen and felt as hallowed ground, and they admonish us that our kind is wont to forget this simple and important fact.

This speech has been passed down to us over a long, convoluted journey lasting more than a century. The voyage has many of the qualities of an oral tradition in that we are not quite sure who said what along the long path these words have taken. Like all good stories, they continue to be changed, adjusted, rearranged as they are told and retold again. So whose words are they?

Here, briefly, is the lowdown. (At the end of this chapter, you'll find a list of sources for more details.) In 1854, Seattle (more correctly, Seathl) made his speech to Isaac Stevens, Commissioner of Indian Affairs for the new Washington Territories, expressing his wish to cede his land peacefully to the government while affirming deep reservations and sadness for the fate of the land and the differences between his culture and ours. Seathl spoke in Lushotseed, his native tongue, and it was translated by Dr. Henry Smith, a young physician who knew the language. Smith realized that the speech was something special and that the gravity of its message was certainly watered down in translation. He is said to have visited Seathl many times over the next several decades to discuss the speech so he could get it down as accurately as possible in English.

Thirty-three years later, this event had already passed into history. Smith published his version of the speech in the *Seattle Sunday Star* in 1887. The style is ornate and Victorian, typical of the way English was supposed to be written at the time, more by people like Smith than Seathl.

For about eighty years, the speech lay hidden in obscurity. The late William Arrowsmith, professor of classics at the University of Texas, discovered Smith's article and decided to reedit it. He was either trying for a more authentic Seathl-like style of writing closer to the way natives would have spoken, or else he was modernizing the words to fit the rebellious spirit of the sixties. You may read and decide for yourself. He published his "translation" in *Arion* in 1969.

Arrowsmith read his speech at a large student gathering on the very first Earth Day, April 1970. Thousands were listening. Among them was Ted Perry, a young film professor who had been hired by the Southern Baptist Television Commission to write a script for a film called *Home* about pollution and the state of the planet. He immediately had an idea: adapt the words of Seathl for our own time, taking the solemnity and grace of Seathl's way of responding to crisis and applying them to the environmental problems America was then facing. This was a bit of a historical fiction, because the original speech is more about the folly of claiming to own the land and the white culture's lack of respect for ancestral ground than it is about the poisoning of the planet by human indifference. Perry proposed to take Seathl, bring him into our world, and imagine—what would he say?

He had no idea how successful his script would be. Once it was shown on network television, the words spread like wildfire around the world. Eighteen thousand people wrote in for copies of the speech. The Southern Baptists sent out a flyer with the text, claiming it really *was* a speech given by Chief Seathl. This is where the lie or the myth began.

Environmental Action magazine published the text in November 1972, this time claiming it was not a speech but a *letter* from Seathl to President Franklin Pierce! Shortly afterward, Northwest Orient Airlines' magazine *Passages* published the "letter," again with no reference to Perry. The speech was published in the *Catholic Herald* in Britain and spread farther around the globe by the World Council of Churches. Monsignor Bruce Kent called the speech "almost a Fifth Gospel."

In 1991, illustrator Susan Jeffers brought out a children's book, *Brother Eagle, Sister Sky: A Message from Chief Seattle,* which sold more than four hundred thousand copies and is found in elementary schools across the country. Her Chief Seattle is a distinguished-looking elder in a headdress with buffalo horns. Indians ride ponies through the prairies and stand in tears before clearcuts. Never mind that Chief Seathl lived in the forests of Puget Sound and never saw a buffalo in his life. He would never have ridden a pony.

In the multicultural nineties, one can't get away with this kind of historical inaccuracy. The story of the fabrication of the speech of Chief Seattle made the front page of the *New York Times* in April 1992: Famous Indian speech turns out to be a fraud! Is it indeed a forgery, or is it a case of the malleability of oral history?

Since that day, Ted Perry's phone has not stopped ringing. Susan Jeffers still won't accept that he had a hand in the words of Seattle because her publisher is sure he's only interested in money: "I can't say he wrote them, because I don't know." Perry insists that he has no interest in payment for the proliferation of his words, only correct attribution of the story, however convoluted its history may be.

I had heard that Perry was sick of people bothering him for the real story of Chief Seattle. After a distinguished teaching career in Texas, he was for a while director of the film department at the Museum of Modern Art. He is now a professor of film and theater at a small college in New England. I had read that he was deeply apologetic and never intended to "steal" the words of Seathl and jury-rig them to address our own environmental problems rather than native land ownership problems. He never intended to write something that has become canonized in the lore of ecological thought. However, it is a great speech, and the greatest version was written by Perry. I wanted to ask Ted not to apologize to the world, which he has done before, but to explain how he now feels to have had a hand in the creation of a timely religious text that transcends creed and authorship to truly belong to the world. Happily, he agreed to be interviewed.

Terra Nova: Your version of Chief Seattle's speech is probably the single most widely known respected, duplicated, and repeated statement of the environmental movement. Millions of copies exist around the world. Did you have any idea that you would create this?

Ted Perry: (*Laughing*)   I would have been more careful.

Terra Nova: This is how religious documents come into being. No one is sure just who wrote them, when, or how. They outlast their creators, and they can't be stopped.

How did you originally want to have the text attributed, when it appeared in the Southern Baptist film *Home?*

Perry: Well, I didn't want to say "based" on Chief Seattle. The original text was just a narration that I wrote. In discussions with the producer, I said I was influenced by and inspired by Bill Arrowsmith's reading of his version at the Earth Day rally. I wanted to give him credit because I called him and asked if I could make use of his work. I said, "Do you care?" and he said, "No, fine." He always honored that, and he did not really have to. It was a verbal understanding we had. And so I did. But without my permission, and without telling me prior to the network airing, the producer decided that it would seem more authentic to take the "written by" credit off. The film then says *researched* by Ted Perry—*words* of Chief Seattle.

I would often protest that I never intended for it to be ascribed to Chief

Seattle, which is true. But I had used his name in the text, which became embarrassing for me when people suggested I had feigned authenticity.

Terra Nova: Once the story came out many years later, have you had people say you were abusing Native American traditions?

Perry: Oh, yes. I have gotten some angry letters. I have gotten more nice letters stuck in envelopes from schoolkids writing to me and saying these were wonderful things I said. But some people felt that they had been betrayed, that they . . . used these words and thought they were authentic and then discovered they were not and blamed me, which I understand.

But I have never received any angry letters from Native Americans. I have heard some have been angry, but they have never written to me. The woman who wrote the *Reader's Digest* story [Mary Murray] went to a school for Native American children in 1993. They actually talked about the Chief Seattle text. The teacher said, "Did Chief Seattle say these words?" The students said no. The teacher said, "Does it matter?" The students said no. I do not know whether these stories are true or not, but I have certainly heard of Native American people who were upset with the whole thing. But it has been used in more crass ways: I think there was a wooden Indian at the 1974 Spokane World's Fair that held out an arm or some- thing and a copy of the Chief Seattle speech came spewing out. Stuff like that.

Terra Nova: Are you familiar with the book called *Mother Earth* by Sam Gill? He argues in that book that the idea that Native Americans worship something called "Mother Earth" was something inspired by a few newspaper articles written by white men in the last century.

Perry: I have heard that argument.

Terra Nova: And that Native Americans picked up on that—it sounded good—and they began to actually follow up on these things. This whole way of dealing with native traditions is an interaction between their culture and ours trying to fit together into something new and unprecedented, a kind of cross-cultural vision that might not be mere stereotype, but instead something to be taken more seriously: different worlds trying to make sense of each other and creating a new language to do it. This does seem appropriate for something as universal as environmental problems. Why should someone be proprietary about ideas that can help here? Or should we be extremely sensitive in appropriating ideas from a culture which we have specifically oppressed?

Perry: I think my attitude about Native Americans in 1970 was very different from my attitude today. It was certainly much easier for me at that time to write a speech for a Native American than I would ever think of doing today. It is always a lot easier to put ideas and words into a mouth of someone else in another culture if you have not really encountered that culture.

Terra Nova: When you see copies of the speech repeated or reprinted, whose version do they resemble most closely?

Perry: Most people seem to feel they can rewrite it to suit their purposes. I am always seeing things that do not match any of the three that are acknowledged. There is a sense that "Gee, this is nice, but what if we change this word here." In the popular children's book, it's all been altered to relate mostly to children.

Terra Nova: Do you think that is fair? Should people be able to rewrite it?

Perry: I did.

Terra Nova: Should it continue to be as flexible?

Perry: When people ask permission from me to publish it, I always have insisted they say, "This is a text *inspired* by Chief Seattle," which is true. I try not to make any claims on it other than that, but I have asked for that. What I object to is people revising it and then still claiming it *is* Chief Seattle's speech.

Terra Nova: I hear excerpts of it all the time. Every month or so, someone says, "I would like to conclude my talk with these words from Chief Seattle." And they are always powerful, but the words always sound a little different, and not everyone who speaks them realizes their convoluted history. Is there a reason for this to keep going on?

Perry: People have to reinvent their mythologies.

Terra Nova: There is obviously a concern for the precision in words, or people would not keep reworking them. There is some attempt to get it right. People want to prove the vision of Seattle is right for their cause.

The case seems very different from *The Education of Little Tree*. That is a very beautiful book about a man who reminisces about his grandparents who were Indians and raised in the traditional way. All these basic gems of wisdom are told in a beautiful, clear simple text. The reputation of this book, too, spread by word of mouth. It made it onto the *Times* bestseller list for months.

The author was Forrest Carter. The book turned out to be a complete fiction. He never had Native grandparents. And worse, Carter under another name was a speech writer for the Ku Klux Klan. All of a sudden, the book was discredited. It's been moved to the fiction shelves. Because Carter is not a moral person anymore, because he was a speech writer for the Klan, we are meant to forget he wrote a wise and inspiring novel. Nevertheless, he has proven he is good with words. He can argue for different causes. It's like rhetoric versus the truth. The book is still a great book. You just cannot presume anything about the author from it anymore. Who knows how much you can presume about the writer of any book?

Perry: It is an interesting lesson. We do not really trust the tale unless we trust the teller. If we can no longer trust the teller, the tale must not be true.

Terra Nova: Of course, there's a whole movement in the study of literature away from that. You only have the text to deal with, but here people are rewriting the texts! The most significant thing, it seems, about the Chief Seattle speech is just how widely known it is, how powerful the words are, and how the story and the idea have just sort of spread around. People want to hear someone from the destroyed culture speak up and say, "This is what is different. This is what is the same." It's larger than life and can't be stopped. Do you feel somewhat haunted by this thing?

Perry: Well, *haunted* is not the right word. I do feel like it will be on my tombstone. Something always appears, or somebody always calls. *Time* magazine called a week or two ago because some politician from Washington State had challenged something that they had printed, and they had concluded with this Chief Seattle text. They did not think the Chief Seattle text was authentic. And this goes on. I remember one particularly angry letter last year when a piece came out in *Newsweek,* but I have not gotten that many. Mostly, they are letters trying to get the story straight and help understand what really happened.

Terra Nova: Now that you know native cultures a little better, would you write this piece differently or just not do it at all? How would you deal with the situation if you were faced with the opportunity today?

Perry: That is a good question. I don't think that I have an answer. I would not immediately discount writing it, but I would certainly like it to be a lot more accurate. I learned much from talking with a Hopi video artist who visited the college last year—the sense in which his people come out of the earth, and they still see themselves in relation to earth, and how different clans are named for different animals. A kind of interdependence with nature and the natural world that still continues, even though he has chosen to interact with the new technology.

Terra Nova: I wonder if putting these powerful admonishing words into the mouth of an Indian chief out of the distant past somehow seems safer than actually saying we want to speak about the earth ourselves. We want to speak poetically, we want to say the earth is sacred, but we are afraid of laughter, retribution. If it's a noble Indian, some respected distant person, then it's somehow more trustworthy and indisputable.

Perry: That's the good side of it. The bad side of it is somehow we have to put these thoughts into something outside of ourselves to give them validity.

Terra Nova: What would happen if people would quote these words without mentioning Chief Seattle? Do you need to bring in the chief to make the message carry more weight, or can any of us just ask "how can you buy and sell the air?" as an American today?

Perry: Well, theoretically you should be able to just ask the questions and state the thoughts, but as we were saying earlier, oftentimes having a particular teller makes the tale. It somehow seems more real.

Terra Nova: A chief besieged by the government. The underdog giving up or making a final plea. But would it be better to write some other speech that would spread just as widely, without appropriating the authority of another time and another culture? Why not accept that these sentiments come from our time and our culture?

Perry: Somebody else should do that.

Terra Nova: Do you think it is worth doing? There are all kinds of environmentalists' proclamations of different types that people are putting out. Declarations of Interdependence, Principles of Sustainable Design, Platforms of Deep Ecology . . .

Perry: I think for these texts to come to have *emotional* impact, they need an *aura* around them.

Terra Nova: They need an aura, all right, but is it fair to the Native American tradition to take their aura so carelessly? If you were Native American, would it be any better if you had made it up? Would you be more allowed to, or is that just the "multi-kulti" mood of the times?

Perry: It is a little like these arguments about the fact that the early Christians had to say that Jesus was born of a virgin in order to justify who he was. Creating that myth was a necessity for the importance of the message.

Terra Nova: Where is the limit to how the speech can be altered? I could imagine people now adding more references to women, minorities, the poor, or farmers to keep the words relevant and timely. It could end up sounding like *Politically Correct Bedtime Stories*. How would that be?

Perry: It would not be pleasing to me. I think that texts grow and evolve not by changes in them, but by changes in the culture that is reading them.

Terra Nova: So should there be some standard, canonized, accepted Chief Seattle speech? Should your version be the standard, with your name as author appearing each time?

Perry: I have never asked for that, nor have I ever taken any money for it except what I was paid for originally.

Terra Nova: But you would like there to be some kind of standard text that should not be changed.

Perry: Oh yes, maybe two texts, one that says "words supposedly spoken by Chief Seattle," and another, "words *inspired* by Chief Seattle."

Terra Nova: Inspired by Chief Seattle? Or written down by Henry Smith, retranslated by William Arrowsmith, and adapted for environmentalism by Ted Perry? A thousand years from now, this speech may still be around, no one will quite know where it came from, with conflicting stories. Remember, there are such things that have been dismissed as being forgeries and then never go away. Like the poetry of Ossian. In the eighteenth century, poems were found in a cave, supposedly ancient, primal words that were then translated to English at the time as these real romantic Hebridean messages of purity. But it turns out one guy had just written them himself and then completely made up the story of how he found them. But they're still in the *Norton Anthology of Poetry* because the *words* were considered—the words are still beautiful.

Words reach people. When people are looking for quotes, they are looking for short things to express the real issue, and your speech does seem to do that. I think it should be accepted as a document of our time: We, you, and enough people have decided that it expresses what we are concerned about right now. It need not be judged for historical accuracy.

Perry: The text can also say something about how our relationship to Native Americans has changed, first since the last century, and then since the 1960s.

Terra Nova: But also standing up for their culture is part of this multicultural age, where people are supposed to do things differently from one another. Environmental problems are something that everyone has to face somehow together. It is kind of mixed. A lot of people I have talked to do tend to want to dismiss this speech and avoid it in their discussions of Native Americans and nature and such, and yet it is still more well known than anything else, reaching more people than so many other more attributable statements because of the striking quality of the words themselves. And so much of environmentalist literature is meant to inspire people. It is judged by how much it moves people, not whether or not it is true.

Perry: Yes, that is the certainly the issue. In some ways, it is less important that it be true than it motivate people to action.

Terra Nova: Nevertheless, in the Chief Seattle speech there are a few real differences between different versions. Did the real Chief Seathl emphasize a real difference between the God of the red man and the God of the white?

Perry: I think the original says, "Our God is not your God," but I took out much of that in my version. There is a portion of the speech in the film, the end of it, that talks about God which was added by the producers of the film. In my speech, I carefully veiled any references to God at all.

Terra Nova: They added that? Bringing the Gods together?

Perry: Well, it was made for Southern Baptists. It had their name on it.

Terra Nova: Rudolf Kaiser reports that the following sentences appear at the end of your version of the speech: "One thing we know. Our God is the same God. This earth is precious to Him. Even the white man cannot be exempt from the common destiny. We may be brothers after all. We shall see." He doesn't seem to know that you didn't write these phrases, that they were added by the film's producers. These words have been repeated in most of the versions of the speech promoted by Christian groups and also appear in the version at the Spokane Expo of 1974. Yet they seem diametrically opposed to sentiments expressed in all earlier versions of the speech.

Perry: I didn't write those words. The producers did, adding them to my original.

Terra Nova: But either it is the same God or a different God, right? How can it be both? Kaiser also reports that Seathl converted to Catholicism in 1830! It all seems rather complex and historically hazy. Many Gods in Hinduism are the same but have different appearances. Conflict or cooperation? Perhaps an oral cul- ture should be flexible, yet it seems that some clear message ought to endure unaltered.

Perry: Well, it seems to me that we always read our own selves—our lives, who we are—into whatever we read. There is a difference between reading your own text into something and then rewriting the text so that it conforms to your own point of view. But I know that in Smith and Arrowsmith's ver- sions the red man's God and the white man's God are *not* the same. Nor are they the same in the text that I wrote and delivered to the Southern Baptists.

Terra Nova: The original Southern Baptist film *Home*—is that still around?

Perry: Yes. There is a copy of it at the local high school. When all this news broke, someone who runs the local AV library sent something to my wife about what a wonderful film it was and how he had used it for years and years and years. It is a film about the pollution of the earth. We shot it in San Francisco and in parts of Manhattan; for instance the Upper West Side had no sewage-treatment system. Just examples of pollution around the country. It seems awfully stiff and dated now—I think in part because the views were so simple, even superficial. None of us was aware of the complexities, the moral issues, even the economic issues that underlay all these problems, or in their solution.

Terra Nova: Yes. There is a book called *The Culture of Nature* by a landscape architect, Alexander Wilson, who talks about changing attitude towards nature in the suburbanization of North America. He talks about nature programs and how these things transformed our views of nature—*Wild Kingdom,* for instance. The book is really good. It has great pictures, great illustrations to show what people aspired toward in the 1940s and '50s.

Today, environmentalism is really being taken to task on many fronts: How much management should be done with nature? How much should the idea of wilderness be the guiding force in conservation? Most people's engagement in nature is not from within a wilderness, but from a human dwelling surrounded by the earth, be it urban, suburban, or rural. Our policy about nature need not just be based on where we go on the weekends. Nobody spoke up for wilderness until civilization was distanced enough from it.

In the Northwest, the forests really appear ravaged. That is the livelihood for people who live there, and there are people who come from elsewhere and say, "Isn't this a shame that clear-cutting is going on." Of course, the forestry practice has to change to keep it going, but the industry will not admit that environmentalists could have a point, and many eco-radicals will argue that forestry just has to go. Environmentalism should probably become less partisan, answering the objections of those with different points of view. It is a debate which includes the Chief Seattle ruckus. People will cite this confusing tale and say, "Look. This just proves that environmentalism is a myth. You people are making up these ideas!" That's just the point. We have made up this myth. We are the people who need this speech, this message, and it is not going to go away. It may well be with us for a thousand years. It will continue to change. Will it still be Chief Seattle speaking to us, or for us?

Perry: I doubt it. People change, and their texts evolve.

Terra Nova: I have one friend who never likes to see anything or read anything that he hears is based on a true story. And I asked him, "Why not, because most people seem especially captivated by anything they suspect really happened?" He replied, "Then you cannot judge the story in itself, as a work of art, because you have this constant standard of what really happened." Which might be stranger than fiction, but not better than fiction. A good story is more than an account of the truth. Our greatest literature is judged on how well the tale was told, not whether or not it really happened.

Perry: I showed the woman from *Reader's Digest* the John Ford film called *The Man Who Shot Liberty Valance.* This man becomes a Senator from a Western state, based on the fact that he was supposed to have killed this outlaw, Liberty Valance, and in fact someone else did. But his whole career is built on this legend. Near the end of the film, he is telling the true story for the first time to a newspaper reporter, and finally the reporter puts down his notebook and says, "I am not going to print this. The legend has become the truth."

Excerpts from Three Versions of Chief Seattle's Speech

Henry Smith's Version
Henry Smith heard Seattle speak in 1854 in his native tongue, Lushotseed, and then recorded it in print in 1887 after visiting Seattle several times to discuss the meaning of the speech.

> Your God seems to us to be partial.
> He came to the white man.
> We never saw Him.
> We never even heard His voice:
> he gave the white man laws
> but He had no word for His red children
> whose teeming millions filled this vast continent
> as the stars fill the firmament.
> No, we are two distinct races
> and must ever remain so.
> There is little in common between us.
> The ashes of our ancestors are sacred
> and their final resting place is hallowed ground,
> while you wander away from the tombs
> of your fathers seemingly without regret.

William Arrowsmith's Version
William Arrowsmith dug up Smith's little-known newspaper article and published a version in 1969, which he later read on the first Earth Day at the University of Texas in 1970.

> Your God is prejudiced.
> He came to the white man.
> We never saw him,
> never even heard his voice.
> He gave the white man laws,
> but he had no word for his red children
> whose numbers once filled this land

as the stars filled the sky.
No, we are two separate races,
and we must stay separate.
There is little in common between us.
To us the ashes of our fathers are sacred.
Their graves are holy ground.
But you are wanderers,
you leave your fathers' graves behind you,
and you do not care.

Ted Perry's Version
Ted Perry wrote his adaptation in 1970 as a narration for a film by the Southern Baptist Television Commission on the environmental crisis.

The white man's god gave him dominion over the beasts, the wood, and
       the red man,
for some special purpose, but that destiny is a mystery to the red man.
We might understand it if we knew what it was the white man dreams,
what hopes he describes to his children on long winter nights,
what visions he burns onto their eyes so that they will wish for tomorrow.
The white man's dreams are hidden from us.
And because they are hidden, we will go our own way. . . .
The white man does not understand.
One portion of land is the same to him as the next,
for he is a wanderer who comes in the night
and borrows from the land whatever he needs.
The earth is not his brother, but his enemy,
and when he has won the struggle,
he moves on.
He leaves his fathers' graves behind, and he does not care.
He kidnaps the earth from his children.
And he does not care.

Henry Smith's Version

Every part of this country is sacred to my people.
Every hill-side, every valley,
every plain and grove

has been hallowed
by some fond memory
or sad experience of my tribe.
Even the rocks
that seem to lie dumb
as they swelter in the sun along the silent seashore
in solemn grandeur
thrill with memories of past events
connected with the fate of my people,
and the very dust under your feet
responds more lovingly to our footsteps than to yours,
because it is the ashes of our ancestors,
and our bare feet are conscious of the sympathetic touch,
for the soil is rich with the life of our kindred.

William Arrowsmith's Version

Every part of this earth is sacred to my people.
Every hillside,
every valley,
every clearing and wood,
is holy in the memory and experience of my people.
Even those unspeaking stones along the shore
are loud with the events and
memories in the life of my people.
The ground beneath your feet responds
more lovingly to our steps than yours,
because it is the ashes of our grandfathers.
Our bare feet know the kindred touch.
The earth is rich with the lives of our kin.

For Further Confusion

Arrowsmith, William. "Speech of Chief Seattle, January 9th, 1855." *Arion* 8 (1969): 461–64.

Carter, Forrest. *The Education of Little Tree.* Albuquerque: University of New Mexico Press, 1990.

Gifford, Eli. *The Many Speeches of Chief Seattle: The Manipulation of Record for Religious, Political, and Environmental Causes.* Rohnert Park, CA: Sonoma State University: Occasional Papers of Native American Studies, no. 1, 1992.

Gifford, Eli, ed. *How Can One Sell the Air? Chief Seattle's Vision.* Summertown, TN: Book Publishing Co., 1992. [Contains complete texts of the Smith, Perry, and Arrowsmith speeches, along with a brief explanation of what happened.]

Gill, Sam. *Mother Earth.* Chicago: University of Chicago Press, 1991.

Jeffers, Susan. *Brother Eagle, Sister Sky: A Message from Chief Seattle.* New York: Penguin, 1991.

Kaiser, Rudolf. "Chief Seattle's Speech(es): American Origins and European Reception." In Brian Swann and Arnold Krupat, eds., *Recovering the Word.* Berkeley: University of California Press, 1987, pp. 497–536.

Kent, Bruce. "A Fifth Gospel." In *Testimony—Chief Seattle.* London: United Society for the Propagation of the Gospel, 1978, pp. 94–98.

Murray, Mary. "Little Green Lie." *Reader's Digest* (July 1993): 100–104.

Smith, Henry. "Scraps from a Diary—Chief Seattle—A Gentleman by Instinct—His Native Eloquence." *Seattle Sunday Star* (Oct. 29, 1887): 10.

Wilson, Alexander. *The Culture of Nature.* Cambridge: Blackwell, 1992.

Light

C. T. Lawrence

In the late 1980s, scientists discovered a hole in the ozone layer. At certain times of the year it appears over the south pole, at other times over Patagonia.

My twelve-year-old daughter sleepwalks in the light, and I, her father, stand watching under the stone-bright afternoon sun. There is too much light. It is everywhere: so strong it makes a black outline around the nightgowned figure my daughter is, turns the grass she walks on a fluorescent green. Her closed-eyed face turns toward me, and it is white, a pebble in a stream.

My daughter says it comes and takes her by the hand. So she climbs to the flat top of a cliff by the sea and lies down in the grass and waits. She says the light comes in clearest up here; I go with her and watch. First she lies down, arms and legs stretched out like she has been flung down by some giant hand, then she stands and tilts her head back, her face lifted to the sky. She says the light would never let her walk off a cliff, but I don't trust it with my only child. I stand between her and the sea, my arms open wide.

Spiderwebs of hair lift off my daughter's head. She tilts it to one side as though she is looking up into a lover's face. And her hand is held out, fingers curled around a handful of air as she walks circles in the wind.

We live in a country at the end of the world. The television shows us maps of the hole in the sky. They say it is coming again. And the next day the light changes, pounds down on us so hard we can feel the heat on the tops of our

Jerry Uelsmann

heads, the backs of our hands. Our shadows stretch at sunset, turning us into stick figures.

The doctors on TV say wear sunglasses, wear hats, don't go outside. The farmers say their sheep are going blind. The drugstores get cases and cases of sunscreen from America. It sells out in a week.

I ask my daughter, who do you see when you go up there? She says there is so much light coming out of his face.

In our country sometimes the sun doesn't set for a month. We crave the cool of our houses so we can go in and bathe our hot, itchy skin. We crave darkness so we can rest our eyes. We want this long day to end. My wife sleeps for eighteen hours and only gets up to light the gas under the kettle when we come home. Inside, everything smells like dust; even our tea tastes dry. We sit at the kitchen table and close all the blinds, the shade cooling us down. Our daughter blinks with her pink-rimmed eyes and, in a voice that sounds like it is someone else's, tells my wife and me what the light has said. Her skin peels and peels. There are circles under her eyes. Neighborhood boys light firecrackers in the street; she speaks to us over far away pops and crackles. My daughter says it tells her that we will be delivered, we are safe, we have nothing at all to fear. The light says that it has a message for the people of my country, and that my daughter is the one to speak. The light says our time of rest is coming: that night will be falling soon.

Adam David Clayman

Power, Protest, and Factory Fumes

Bikram Narayan Nanda and Mohammad Talib

Photographs by Adam David Clayman

[1]

Power oppresses in that it inflicts itself on its subjects without heed to their will. Protest emancipates in that it disapproves of the precepts of power and is able to align itself with the realm of the possible. It is power's immediate imperative to forget or lie about the damages perpetrated on people. Conversely, it is the survival imperative of the people to wake up to count their damages and organize a protracted resistance.

Power perpetuates social amnesia[1] through various cultural modes, forms, and categories. Protest wakes up to remember the sores of the damaged life and its condition through a repertoire of folklore and local memory, collective nostalgia of the forsaken, real experiences and stirred up visions. Thus, social amnesia, which makes the reality of wage and the immense shadow of industrial serfdom possible, forces people to forget and forfeit their real experiences. There are forgotten chapters in the charted path of industrial capitalism. Ecology is but one example of the casualty of this social amnesia. It scarcely enters the calculus of present prosperity and is rarely accounted for.

The oppression due to industrial pollution and the reality of factory fumes in a suburban village in Delhi provide the setting for this account of a power that sabotages real experiences of the victims of toxicity, leading to a protest that refuses to forget.[2]

[2]

We begin by relating a story of a lost traveler, but it is not exactly that of the fairy-tale lost traveler, the ardent disciple of a great guru who, in his search for truth, happened to stray into the labyrinthine blinds of the woods. That original traveler was rescued by the illuminating sight of a distant smoke that signaled a living human settlement. To him, smoke symbolized life. Not only life: It alluded to hope, direction, and meaning.

Smoke was once a metaphor of humanity.

In this sense, it still finds echoes in the contemporary rhetoric of development: "Smoke is progress!" An apocalypse of civilization, the aura of the metaphor continues to invigorate reckless industrialism. It restores to industrialism the authenticity of folklore and the conviction of a legend. It mythologizes the barbaric advances of industrialism and clears space for lodging it. In the same breath, it sanctifies its lethal shadow on displaced peoples and their eroded cultures.

Metaphors of dominance seek to sabotage real human experiences. They refuse to understand, to listen to the anguish of the victim. In the experience of the pain and agony of the victims, such metaphors become a sham, a blatant lie. The horrors of experience question their veracity, and when the firewind of popular scorn begins to blow, it strips the metaphors naked and forces them down on the ground.

Smoke no longer stands for progress.

Now new metaphors are born, and an alternative history is given a name. Cataloged in people's real experiences and compiled into petitions, pamphlets, graffiti, and posters, this new history repudiates the dominant metaphors. Smoke becomes an antithesis of humanity, a negation of life. It connotes damage, despair, disillusionment, and disorder.

In our story, the traveler, to pursue the old fairy tale, is lost not in spite of but because of the smoke. This is not the domestic smoke born in a complacent hearth. It is, on the contrary, an unbridled smoke manufactured in the turbulent cauldron of an ambitious factory. This toxic smoke is belched out of high-temperature digestive chambers of a factory that combusts fossil fuel. The smoke of the hearth symbolizes life; that of this factory alarms it. The former suggests solidarity; the latter evokes dissent. Neither the metaphors nor their logic approximate human reality.

To abandon a metaphor too abruptly is to thwart a deeper understanding of the endless intricacies of social process. Living characters act out the metaphors, sometimes conforming, elsewhere outstepping, in real situations with real consequences.

Smoke as progress has thrown up its own manner of heroes. Smoke as life's negation has its champions too. These are living people with real names, and they have refused to disappear as statistics of industrial disaster. They haven't yet surrendered their memories. They have resisted dominant images that obscure and deny their real experiences. They have battled every day to defend their symbols and meanings from their usurpers. In their stupendous will and energy they have guarded their arduous discoveries. Their struggle is seldom orchestrated by a pristine theory. Their unmitigated protest and protracted struggles have outlived the promptings of political dogma.

In our story, smoke oppresses and does not invite. Smoke converts the lost traveler into a gallant crusader.

[3]

The traveler of our story is a schoolteacher named Bhudutt who inhabits Meethapur,[3] a suburban village on the outskirts of Delhi. He arrived here as one of the many immigrants[4] from the agricultural hinterland surrounding the expanding metropolis. Before his encounter with the factory smoke—toxic fumes emitted from the neighboring Bitumen Purification Plant—Bhudutt had already traveled a long winding biographical trajectory.

He was born in a "natural" world: a world peopled by peasants who eked out their existence in a "sacred" environment. Innumerable causes of concern and anxiety beset the people, but never the freshness of air and the purity of water. That fresh air and pure water could be scarce never occurred to Bhudutt in his ancestral village. As a child riding on a buffalo in the pastures, he used to roll in the exuberance of an afternoon breeze. The experience always exalted him. Running helter-skelter in the open and swimming in the nearby river with his village mates were his most coveted escapades. The smoke in the atmosphere invariably drew its stuff from the *chulhas* (hearth) in the *rasois* (kitchen). To Bhudutt, smoke was a natural associate of cooking food. Its sight reinforced the everyday faith that his hungry stomach, after the day's exhausting work and wandering, would surely be satiated toward the end. Smoke meant food. It was as familiar to him as the clouds or the fog. There were little variations in its color or character. This was the symbolic patrimony that Bhudutt had come to inherit from his ancestral past, in which environment was taken for granted.

Apart from that patrimony, he had acquired an amalgam of aspirations through his modern education that could not have been locally realized. Bhudutt's ideal life, ironically, lay elsewhere. In his search, Bhudutt had to bid farewell to his past association, but the umbilical was not entirely severed. His present could scarcely bypass his past, let alone transcend it. Bhudutt's journey, his physical displacement, paradoxically paved the way for an emotional journey into his past. At Meethapur, his new destination, he had no other symbolic anchorage except his past. In a sense, his past armed Bhudutt with a certain inner resilience to cope with the rugged imperatives of the present. Where the present was anarchic, the past was cathartic.

When the past descended on Bhudutt, it did not seek his concurrence. This is not to say that it was his personal prerogative to say no. The deluge of the past was irresistible. The deluge deposited the forsaken rudiments of his past in the innermost recess of the mind. *That* was nostalgia.

Vivid reminiscences of associations and participation, festivities and quarrels, gaieties and anxieties of the village got mysteriously forged together to evolve into a bizarre bewitching cloud-nostalgia. Nostalgia was a quiet cry of the abandoned biographical past which now lay animated in the suspensions of personal memories. Like a lost soul or a desperate hope, nostalgia possessed Bhudutt. Nostalgia was Bhudutt's personal historian who recorded biographical travelogues on the sheets of memory. Like a usurious bookkeeper, nostalgia always placed Bhudutt in arrears. Nostalgia was not only the past's reincarnation, but its litany as well. In Bhudutt's mind, his fascination with the past did not derive its nourishment from totally unwarranted grounds. It was basically rooted in the moist soils of partial memories, of trust and mutual help, of enduring charcter and unflinching confidence. Nostalgia turned the past into all harmony and order.

Had it not been for the force of circumstances, Bhudutt reiterated firmly, he could not have budged an inch from his ancestral village. But the reality of the present was more real than the reality of the past. The vast aggregation of everyday concerns of the mundane present created a barrier reef of their own. Nostalgia, in which he sailed into his past, ran aground on such a reef just short of being shipwrecked. Thus Bhudutt was thrown overboard onto that reef: a totality of numerous routine matters for eking out subsistence. This terrain was to provide him a lasting abode.

No doubt he built a house for himself, but it was not until he got a teaching job in a local school that his house was roofed. The roof adorned his name with local respectability. He relentlessly mobilized the material resources available to him in order to earn moral credibility, mostly occupying himself after school hours with the education of not only his own children, but also his neighbor's. Bhudutt's immigrant status made him more vulnerable to the imperatives of his present. Often enough, he suffered from the anxiety of falling out from the collective rhythms and local aspirations. Thus he could neither afford to withdraw into himself nor into his memories. Everything he did, without exception, thrust him into accepting and adopting the new rhythms of his chosen life. Bhudutt ought to have lived happily thereafter, but this was not to be.

Just when he had carved out a protective lair in the dense coral reef of everyday concerns, a cataclysmic continental drift impaired the organic ligament of his innocent shell. The impetuous eye of the drift was the Bitumen Purification Plant.[5] The location of the factory on the outskirts of the village was, in reality, the dislocation of his life's bone out of its socket. On the inauguration day of the factory, Bhudutt was unaware of the lethal tremors that would one day unrock the reserves of his life and symbols. These lethal tremors were none other than the noxious fumes that billowed out of the haunting chimneys of the factory.

Our traveler was lost once again.

At first, his enduring politeness refused to admit the physical discomfort of the vile yoke of recurrent fumes. A beleaguered optimism nurtured in him an unquestioned faith: After all, the village sky was too immense for the fumes to pollute it. Very soon, however, the imposing spectral face of the fumes began to sicken Bhudutt's gut. The sickness permeated every moment of his daily existence—a daily torture for his senses. He watched in utter anguish the fever-stricken faces of his wife and children, which mirrored the grotesque fumes. He perceived the fumes with agonizing distinctness. During the day, they recurred an infinity of times with a taunting inevitability, returning ever more insidiously, violently, and defiantly.

On the face of a cloud of ravishing whiteness, a raven cloud of fumes extricated itself and hung with all the details of a monstrously complicated molding. On Meethapur's rainwashed sky, the daily repetitions of the fumes formed a canopy of sickness out of the broken and blurred limbs of Bhudutt's soul, which moaned where everything of the original life was threatened. Day in, day out, the fever burned unabated. In a burst of anguish (or was it deliberate anger?), Bhudutt coughed out the politeness that had congested his bronchial experiences of the past days. When others coughed, Bhudutt realized that the fever had burned them too. Now his experience of fumes—the unwilling captive of personal politeness and etiquette—was decaged. He had realized the impossibility of living face-to-face with the reality of the fumes, but he was no more a solitary insomniac of the day bed nor a lonely homeless soul. In losing his way, he had found himself in the midst of other travelers who were similarly lost.

The villagers, without moving anywhere, had become migrants in their own homes. Unlike his earlier migration, Bhudutt's present one was not willed. Moreover, it was a migration in the inverse. What had really migrated, or rather was banished, was Meethapur's environment. What had remained of it was nostalgia, which swept the shores of local memory.

This time it rode the foaming tide of collective scorn. Earlier, Bhudutt lamented his ancestral past. Now he sorrowfully harkened back to the intimate memory of Meethapur's forsaken environment. The not-too-distant past now evoked in him a certain blissful feeling of exceptional purity. He had overcome his longing for his ancestral village, but the memory of Meethapur's past environment was to stay with him, within him. It stuck like the silver sand of the river to the skin of his soles, lived in his eyes, on his nerve tips, and gave depth and substance to the background of life's everyday hopes.

Some days, interrupting his routine, he thought he would look through the window and see Meethapur's spring untainted with the factory fumes. Earlier, the ship of his lonely nostalgia had wrecked on the coral reef of everyday concerns. That coral reef was now objectified into a floating island of collective nostalgia. Despite the fierce fondness for the past that nostalgia evoked in him, everything he had imagined with pictorial clarity paled, crumbled, and withered in the face of the tormenting fumes. As in a nightmare the caravan disappears behind the back of the marooned gypsy, so did Bhudutt's past disappear—a farewell twitter of birds and children's rhymes in a configuration of trees, meadow, and breeze. Everything in raw memory seemed to collapse like a scrap of rainbow faded in the industrial sky or like the *jhuggis* (encampments) swept away by the State's men while the people had gone to see someone off.

Thus, nostalgia was only visibly an affair with the past, but in its not-so-visible yet harder moments it was a struggle with the present. In the dialectic of these two moments, nostalgia stepped straight out of its passive submission to reminiscence into an active retrospective review of the past. This enabled Bhudutt to comprehend and visualize the complex prospect of where he had come from. He could extract conviction and courage to resolutely confront the predicament of the present. He could easily derive that the factory was not inevitable, nor was the polluted atmosphere unredeemable. Nothing about it was either indelible or bearable, and he decided at once that he would not take it anymore. To Bhudutt, the factory was a ghost with neither a name nor a home, and it must eventually fade away. The factory ought not to exist.

Thus nostalgia informed and fed on an analytic edge of the mind that thrust him forward into the future. It is in this sense that the present nostalgia was a sword-swish of a political sentiment, a defying stroke of a tense-willed mind.

In his nostalgia, Bhudutt was living his past backward into the future. The experience of the fresh environment of yesteryears never occurred to his mind before. The experience of fresh air curiously entered into his sharp mental focus, through the smoke screen of the foul fumes. He was somehow lifted above the desperate quagmire of troubles and privations, so that he could see all sorts of distant but overwhelming possibilities, just as, when a little boy, his mother used to lift him by his elbows, enabling him to see what was mysteriously hidden beyond an impossible fence. Bhudutt's existence was the dream of the vision of an environment he had perceived, or appeared to perceive, in his own dream.

[4]

Bhudutt now had a solid realization that the fumes could be warded off through active deliberations. There were animated silences and discomforting blanks that punctuated the political trajectory from a conception of the possibility to the vivid awareness of practical difficulties entailed in its actualization.

Bhudutt was not alone in that long travail. There were many more. For the moment, at least, there were two others: Maasterji (another schoolteacher of the village, middle-aged) and Punditji (a tenant farmer in his early fifties) were coparceners to the common convulsions and realizations of their collective predicament.

This instance takes our narrative to the drearily drenched monsoon night, Bhudutt's living room, July 28, 1984. The trio was so engrossed in making laborious notes in cipher of their laborious inspections and speculations that they were unaware of their wives' agonized efforts to burn the damp firewood without flame, of the screams of the teething baby, abandoned on the floor, of the broken croons of a shivering puppy dying in the adjoining alley. What of the fumes and the factory? These were etched in the scowling language of despair, the sense of loss, the awareness of sordid deceptions, a dizzy feeling of injustice, and the helplessness of the wasted lungs. All this almost drove them out of their minds, but not quite.

The dark of hopelessness and servitude was not to stay with them for long. It was dispelled like the night at the day's awakening—painfully slow in the beginning but suddenly dispersing into an immensely illuminating release of energy. A painstaking fate had put the three of them together, forming a single shadow on the wall that wavered at the mercy of the flickering candle in the lizardy corner of the room. Indeed, the shadow symbolized the dialectics of hope and despair in the trio. The shadow was the burning candle's primeval pledge to the day to keep away the night until the dawn. It was the fluid image of the luminous possibility at night. Although its texture resembled the dark, its essence presumed the light. A constant companion to the glowing candle, the shadow breathed its life from the radiating rays. It was a unified spread of the trio in livid confrontation with the night. The shadow was a shadow of persistent witnesses to history. It was a measure of something not quite comprehensible in the trio, but overwhelming and ingenious to be able to transcend its surroundings.

The quiet quivers of the shadow were not merely mirroring the whimsical flickering of the candlelight. After all, this was the shadow of the spasms and jerks of an intensely debating trio breaking into intermittent gestures: defiant fists, disapproving heads, and agitated limbs.

The late night hour was heavy with an involved seriousness that settled over a sorrowful archive—a dossier. Out of swarms of carbon copies, photocopies, endless extracts from pocket books on law, illegible jottings on stray pieces of paper, ballpoint remarks straggling over the margins of their own handwritings; out of half-crossed-out paragraphs, unfinished sentences, and improvidently abbreviated though not yet forgotten names, dates, and places—out of all this they were making and remaking a lucid, orderly dossier on an otherwise disorderly factory and its fumes.

All the fire and fury, all the splinters of bursting passion, all the insistent strength of will and mind, everything that was supposed to erupt at the point of confrontation, to burst forth and clutch in its hold even if only for a short time, the subaltern power, to shake violently the shackle, and perhaps to crimson the village sky—all this now found its expression in the froth and ferment of the dossier. One can say, in fact, that here was the aim and ardor of the dialectic of despair and hope that had long been accumulating in muffled depths: These confident and disquieting letters to the custodians of law and industrial order were all stacked in place.

The dossier formed an ignition point in a relentless struggle against the factory fumes. In writing their material, the trio resembled seasoned inventors in a science workshop employing an infallible ability to fasten, to attach, to solder inert facets of their scattered speculations and reflections. Right through the night the trio brooded over their encounters and experiences, sometimes with involved discreetness, sometimes with a casual frankness interspersed with moving shades and dispersions of a flickering flame.

Toward late night, the trio began to show signs of fatigue. By the time the discussion had tapered to an end, the candle began to die out. There was a sudden flash of illumination, and then all was dark. The shadow on the wall of course lost its distinct circumference to dissolve eventually into a larger collectivity. Their bodies sweated in the dilated joy of giving birth to a movement. These were not the moments of absolute nightmare, the abstraction to end all abstractions, an infinity pitched against an infinity. In reality, it was the first resolute step in the long journey of a thousand miles.

[5]

For Bhudutt, the fumes had damaged the field of the given. He was able to take a glimpse of the possible with more than the usual state of emotion, with more than the usual grasp of the ought-not-to-be, under the spell of moral anger that was deep and vehement. This encounter of the possible could be ordered in Bhudutt under the shock of the oppression of factory fumes: the possible that was otherwise suppressed or never recognized under the inertia of routine and familiarity.

These moments are peculiarly vulnerable to appropriations by the prevailing *doxa,* only to be snubbed and taunted as "wild spasms," "unbridled spontaneities," and "idle speculations." Even the discourse of protest literature somehow

Adam David Clayman

disavows rare existential moments in the experience of the possible. This disbelief stems from the intangibility of the moment.

[6]

Our sketch of protest has not yet touched on power. It is natural to expect that the turmoil so characteristic of the episode should have deeply affected the capitalist, the man behind the factory in question. The predominant image of the capitalist projected by the villagers should have jolted any pretension to morality or entrepreneurial integrity.

Unexpectedly, interviews and observations during our fieldwork revealed an unruffled poise and a remarkable grace in the capitalist. Unlike the affected villagers, he was ostensibly cheerful and calm. The nature of the reactions by the capitalist was diametrically opposed to that of the villagers. Whereas the villagers were publicly abusive and personally vile in their reactions, the capitalist was sardonically detached and recklessly prudent. The villagers had personal scores to settle, but the capitalist had a "mission" to accomplish. The villagers hurled slanders; the capitalist advanced arguments.

In the worldview of the capitalist, his mission had a divinity of its own. This divinity drew its aura from the divine propensity of his material capital to beget itself and multiply. This task of augmenting capital, according to the capitalist, conferred both honor and credibility on his enterprise. The internal logic of this divine quest was simple. Whatever promoted the proliferation of his enterprise was moral, and the obstacles in his way were outrightly immoral.

The capitalist's worldview glorified the *self* and degraded the *other*—in this case, the villagers and their world. The self constituted the domain of the moral and the humane. The other was seen as the loathsome counterpoint to the self: a Machiavellian sphere of no principles, of flagrant dishonesty.

The self, however, did not assign to itself an unqualified righteousness. It knew for sure that the world of impiety was ever so contagious. Had it not been for the commitment to the mission, the self could hardly have remained uncontaminated, the conscience so clear, the being so steadfast. The self, in its becoming, could indeed strike a delicate balance that instilled an emotional equanimity in its tenor. Having acquired grace, the self neither overdefined the other—the protesting enemy—nor undermined its inner moral strength.

Such a strength provided a remarkable fearlessness. It was this ability of the self, our capitalist assured, that could prevent him from mistaking the silhouette of a tree for a bandit. He was fully aware of the hazards of inner fear and the damages it inflicted on the self. To him, fear imposed its own reality. With fear in the eye, one was liable to encounter a scorpion in every pit, a lion in every jungle, and a thug on every roadside. If fear rendered reality formidable, then fearlessness made it amicable. Fearlessness had the magical power of turning the most venomous of snakes into an innocuous rope—the self that cast its shadow onto the reality.

Equipped with a fearless self, the capitalist regarded the villagers' threats as nothing more than intemperate moods and empty phrases. Surely, he had little reason to be swayed into losing his sense of reality. He harbored in his mind the devious craft to contain symbolically the villagers' resenting threats. Our capitalist's grace was thus nurtured in the resolute confidence of a self-validating project.

There was no doubt in his mind that history provided innumerable instances of the momentary persecution of the good by the evil. He commandingly cited a repertoire of historical events and religious legends to support his contention. Be it history or myth, our capitalist related the anecdotes as allegories, a metaphorical medium to organize the self vis-à-vis the other.

Here was the self in search of a parable, a legitimacy, a convincing lore. In this search, history was a resource to be appropriated. The capitalist could recount, "Christ was crucified by the heathens. Muhammad was tortured by the Kafirs (infidels). Rama was challenged by the Asuras (demons). Were their predicaments any different from mine?"

In the worldview of the capitalist, the villagers were a necessary evil. Their conduct and dispositions did not belong to the moral terrain of life. Thus the inclusion of the capitalist's factory into the ambit of morality had to exclude the villagers. The fumes were emitted from within the moral domain. They were therefore acceptable as part of the good. Because the villagers belonged to the immoral domain, their credibility to protest against the fumes was indeed suspect. If and when they complained, it only reflected yet another instance of immorality.

The village head, the Pradhan, never authored any letters of protest to the state officials. In fact, he countered specific allegations of the protesting villagers by writing letters favoring the capitalist. With the moral support of the Pradhan, the capitalist could now disparagingly declare that protest was an unfortunate outcome of village factionalism. He attributed this factionalism to two antagonistic groups composed of those who rallied around the elected Pradhan and those who connived with the defeated candidate. The factory pollution was a mere pretext for the resenting faction. In the capitalist's metaphor, it was "a shoulder to cry on." The local factional rivalries and the personality clashes had concocted an issue out of the fumes and turned the capitalist into an ineluctable victim.

The capitalist was convinced that the local villagers were gullible and vulnerable to stake their self-esteem at the slightest lure of material gain. They lacked a sense of decorum, and they had little respect for a proper work ethic. He attributed the villagers' resentment over the fumes to the fact that they had nothing worthwhile to do. Even though many of the villagers were employed in the neighboring industrial estate, he had serious reasons to question their integrity at work. In a contemptuous voice, he pointed out that they punched their attendance cards at the factory gate and returned home unafraid and without compunction. He warned us not to trust the villagers for their words. He categorically dismissed their complaints against the factory fumes as blatant lies. As a matter of fact, the factory fumes were scapegoats to extract favors from the capitalist. He made it clear that the villagers were never more cheerful than when he satiated their vices. He could effortlessly count the names of individual villagers on his fingertips who owed him bottles of alcohol, TV and receiver sets, bags of cement, liberal loans that would never be repaid, and countless other gifts on innumerable occasions. In his view, most villagers were chronic alcoholics lacking in reliability. He had no doubt in his mind that a bottle of alcohol every evening is all that the most vocal of the resenting villagers wanted from him: "If instead of water, alcohol was to flow in the canal, the grievances against the factory fumes would be once and for all settled."

The steady disintegration of the moral community of the village and the subsequent decline of the traditional norms and expectations provided the capitalist a convenient ground to discredit the villagers. According to him, the villagers were violating their own standards of morality. Their notorious preference for money to morality accounted for much of the social malaise in their everyday life. He alleged that the men in the village were imprudently promiscuous, even at the cost of the honor of their women. He expressed a curious concern for the awful plight of upper-caste parents whose daughters and sons were unable to receive proper education and became unfortunate victims of sex and lust.

To our query, "How do the villagers reciprocate your favors?" our capitalist reserved a categorical answer. He dismissed the idea that he ever harbored any expectation from the villagers in return for his gifts. Ironically, he dreaded the day when the villagers would stand to reciprocate on equal terms. In fact he implored, "Oh God, do not give them strength to pay me back. Modesty and shame would not let them face up." To our subsequent query as to the reason for being ever so persistent in his benevolence, the capitalist in his characteristic wit narrated to us the following tale of a *rishi* (sage):

> Once a *rishi,* while bathing in a river, saw a scorpion drowning. In an effort to rescue it, the *rishi* impulsively held it in his palm. The scorpion stung him instantly. The *rishi* had to withdraw his hand with a sudden jerk. The scorpion fell back in the water. Once again, the *rishi* reached out to save the scorpion. Again the scorpion stung him. The *rishi's* hand jerked, and again the scorpion fell back in the water. This was repeated several times over. The *rishi* stubbornly insisted on rescuing the scorpion, but the scorpion stuck to his biting instinct.
>
> Eventually, however, the *rishi* managed to succeed. The scorpion was rescued. By then, its store of poison, after repeated bites, was exhausted. On his return from the riverbank, his fellow saints ridiculed him upon hearing the story. Clearly, the *rishi* acted as a fool to have saved the ungrateful scorpion. The *rishi,* however, was steadfast. He had his own reasons to have acted as he did. He explained, "If the scorpion does not want to part with its bad deeds, why should I part with my good ones? I know for sure that bad deeds are always limited, while unlimited good prevails in the end."

Our capitalist likened himself to the *rishi* in the story and compared the villagers to the scorpion. His endeavors to continue with factory production and "alleviate" the condition of life in the village (through the various material favors) were good deeds that he was unwilling to give up because he knew that in the end, his factory would prevail. He made no bones about the fact that he had tolerated the abuses hurled at him by the villagers like a sage. He reiterated, "If the villagers are bad, it does not mean that I too should be bad or lose my patience." The behavior sparked by the factory fumes was summarily dismissed by the determined capitalist as the "scorpion bites" in the story. He was ever so confident that the stock of poison would exhaust itself one day.

Our capitalist was convinced that the villagers' resentment would be short-lived. He compared his present predicament to that of a passionate traveler in the midst of a whirlwind, but he cherished a wisdom from one *tagore* (this was not Tagore with a capital T):

> If the wind is blowing fast,
> Do not face it,
> Just sit down;
> The wind will blow past.

Yet there was a style in which one must steer oneself through rough weather. He had fully understood the futility of fighting against strong currents. His logic dictated to him the strategy of negotiating hostile situations. In an instructive way he pleaded, "If you want to cross the river, do not cross it in the upward stream, but flow according to the flow of the stream [for the time being]." In his perception of the self, through this long detour, the capitalist considered himself neither solely victorious nor indeed vanquished. The episode was too trivial to him to be assessed in these idioms at all.

In every single case, right from the beginning, he stood above the "petty exchanges" of challenge and riposte. He had come to perceive the villagers as a store of venom. However, the village itself sans the resenting villagers was extolled by him as his own body. In this sense, his body was Meethapur, whose life was perpetually threatened by the venom of the protesting villagers: "Only digest them [the resenting villagers], and my body would be cleansed."

[7]

In this part of the story, the capitalist's enduring grace might be mistaken for an unfailing attribute intrinsic to his class. In our own experience, during the initial period of contact with the capitalist, we were summarily dismissive of this grace as truly unrealistic. Most striking was its contrived nature. This was partly reinforced by our ideological predilection, which outrightly rejected any pretension to morality on the part of the capitalist. For us, this morality was a necessary facade beneath which his real interests were hidden, but our bias obstructed us from deciphering the capitalist's morality. We were soon to recognize that the capitalist's grace in reality was neither a virtue nor a vice. It was, if anything, a strategy, a practical necessity pretending to be an eternal virtue.

Such a strategy concealed the truth of its own practice from itself. In so doing, it censored the moments of uncertainties, ambiguities, and anxieties from the real situation. This expelled time that was essentially irretrievable. The capitalist, then, in this detemporalized space was hardly tormented by time. Time, as far as he was concerned, was on his side. He never erred or slipped into clumsy mannerisms. There were no awkward moves nor any amateurish lack of etiquette. In this way he had turned the reality of practical conduct into a ruse.

In such a restricted scenario, the capitalist not only possessed an ability to guide his own conduct, but also mastered the practical art of anticipating untold situations. Even before the protesting villagers posed uncomfortable questions, the capitalist was ready with befitting answers. He was seldom a victim of the trauma associated with the uncertainties of the forthcoming. Even the villagers' actions, however hostile and revengeful in all their numerous consequences, could never digress from the boundaries of the predictable. The protesting villagers could hardly take the capitalist unawares. Even when the capitalist did not anticipate a rebuke here and a slander there, he was ever convinced that the situation was, after all, in his favor. He had mastered the art of the retrospective maneuvering of rules and meanings given to situations.

We had little access into the realities and realms of the capitalist's private and secret exchanges with the state officials and the co-opted villagers. It became increasingly clear to us that there were zones of secrecy that clouded such exchanges. Whereas entry into these zones was possible at the periphery, there was no way in which we could secure entry into the more intimate but significant discourses involving the capitalist, the obliging officials, and the one staunch supporter among the villagers. There were inherent hazards in gaining access to these spheres. To befriend the capitalist implied a loss of our credibility in the eyes of the resenting villagers. In such a conflicting situation, our inclusion in one domain meant exclusion from others.

The capitalist's grace was strikingly noticeable because we had not seen it as contingent on the political strength of the protest against the factory pollution. Our thesis of the grace was not informed of its antithesis. Somehow, we were intrigued by the capitalist's grace, which remained unnerved in the face of popular protest. That this popular protest was fledgling at that time became clear to us only as the capitalist's grace began to show signs of wilt. While the capitalist was still pretending to be graceful and composed, there appeared visible traces of fear and anxiety in his demeanor.

As the facade of grace began to crumble, the underlying practical strategy, which had been hidden, came to the surface. With the passage of time, as the protest turned formidable, the capitalist lapsed into moments of despair and self-dissolution. A deep agony marred his poise as he reflected on his decision to have opted for Meethapur in the first place: "*Yeh to Meethapur Nahin Karvapur hai*" (It is not a sweet but a bitter abode).

Ironically, the capitalist confessed sleeplessness through the nights. This was not caused by the fumes. Now he was faced with the ordeal of endless visits to the courts, government offices, and important people who mattered in the affairs of the state. He had to plead more than the innocuousness of his actions. From the very beginning, the capitalist preferred to remain alone in his venture. He had never involved the members of his family, nor did he share with them his tribulation at Meethapur. Inevitably, he became a lone sufferer. Curiously, he felt relieved over the fact that at least his family remained out of the "filth" he was in. He never initiated his able son into the "mission" (factory), nor did he intend to do so: "*Itni Jaldi usko is Gandh Main kyon Dalen?*" (Why dirty his life at such an early age?) The capitalist had begun to loathe his entire venture.

Yet in the most trying moments amid intense hostilities, the capitalist, while feeling loathsome, was somehow rescued by his grace. Even when his grace seemed to wither away, it never really deserted him. Its apparent fragility was, if anything, the nervous moments of an otherwise stubbornly enduring grace. On more than one occasion, under spells of anguish, the capitalist had even contemplated quitting Meethapur. Our query about this "bitter" option during a long conversation abruptly resurrected the lost sage in him. He resourcefully expounded, "God is everywhere, here and there, Meethapur and elsewhere. Then where is the need to quit Meethapur? I can realize Him here."

This, then, was the dialectic of moral grace and practical strategy: two diametrically opposed tensions within the capitalist. As it happened, when the facade of grace crumbled and the capitalist was rendered defenseless, he invariably transposed his practical strategies as retrospective necessities. He could shift the moral responsibility from his own person to others or to forces of circumstance. In this way, he could justifiably reduce his intended doing into preordained destiny. As an illustration, the factory itself was entirely the outcome of his personal endeavor, yet it was projected as a mission, whose chosen guardian to execute it was no one else but the capitalist himself.

When the fable of preordained destiny did not warrant his practice, he readily resorted to the discourse of moral grace. Between wise sayings and rare maxims of moral grace in a practical world, the capitalist wrestled to resolve the contradictory tensions within him.

The tension between grace and strategy resided within the capitalist, but it was created by protracted protests by the resenting villagers. Sometimes gracefully, sometimes clumsily, the capitalist harbored this opposing tension without ever abandoning his grace or surrendering his acute sense of practical strategy. This dialectic between grace and strategy had made it possible for the capitalist to contend with local protest, and yet it was the local protest itself that contained within it the key to the possible resolution of the capitalist's inner tension. Whether the capitalist will give up his grace once and for all, stripped of the pretentious face, or whether he will foresake his strategy and become true to his grace is not for us to say. In fact, if this dialectic is ever resolved, it will indeed bear the imprints of the logic of practical struggle of Bhudutt and the other villagers in their everyday resistance.

Notes

1. "Social amnesia is society's repression of remembrance—Society's own past. It is a psychic commodity of the Commodity Society." See Russell Jacoby, *Social Amnesia* (Boston: Beacon, 1975).

2. In Jacoby's formulation, social amnesia can be transcended by remembering what it forgets.

3. Before 1947, Meethapur was predominantly inhabited by Muslim peasants. In the wake of the country's partition, the communal riots that broke out willy-nilly forced a large number of households to migrate over to Pakistan. Within the following decade, high-caste Hindus accounted for the majority of the land-owning households in the newly emerged settlement. In the course of time, the settlement itself became vulnerable to encroachment by outsiders. These outsiders, most of them neighboring Gujars (traditionally pastoralists), allegedly indulged in theft and "antisocial" activities, causing commotion among the Brahmins and the village elders. With the moving out of Caste Hindus a significant gap was created in the village social space. There was a real need now for a "warrior caste" to protect the village, both territorially and socially. In this situation, who could better protect than the intruders themselves? These original intruders were eventually assimilated into the village.

4. There were two major attractions for the new settlers coming to Meethapur from the rural hinterland. First, the emerging industrial belt provided job opportunities for the migrants. A large number of the villagers are employed at the neighboring Badarpur Thermal Power Station. Second, for these migrants Meethapur resembled their native environment. It provided the warmth and symbols of their rural past together with the supposed material comforts of an urban town. These new settlers came to be known as outsiders. The territorial boundary of Meethapur gradually increased.

5. In the winter of 1983, the foundation of Hindustan Petrochemicals Pvt. Ltd., was laid. An ambitious entrepreneur guided by his calculus of costs and benefits selected Meethapur as the locale of his factory. This village provided low-cost land and labor as well as an already developed road network. These were important incentives for the capitalist to select Meethapur for his 30,000,000-rupee investment.

Adam David Chayman

Being Prey

Val Plumwood

In the early wet season, Kakadu's paperbark wetlands are especially stunning, as the water lilies weave white, pink, and blue patterns of dreamlike beauty over the shining towers of thundercloud reflected in their still waters. Yesterday, the water lilies and the wonderful bird life had enticed me into a joyous afternoon's idyll as I ventured onto the East Alligator Lagoon for the first time in a canoe lent by the park service. "You can play about on the backwaters," the ranger had said, "but don't go onto the main river channel. The current's too swift, and if you get into trouble, there are the crocodiles. Lots of them along the river!" I followed his advice carefully and glutted myself on the magical beauty and bird life of the lily lagoons, untroubled by crocodiles.

Today, I was tempted to repeat that wonderful experience despite the light drizzle beginning to fall as I made my way to the canoe launch site. I set off on a day trip in search of an Aboriginal rock art site across the lagoon and up a side channel. The drizzle turned to rain within a few hours, and the magic was lost. At thirty-five degrees Celsius, the wet season rains could be experienced as comfortable and welcome: They were late this year, and the parched land and all of its inhabitants eagerly awaited their relief. Today in the rain, though, the birds were invisible, the water lilies were sparser, and the lagoon seemed even a little menacing. I noticed now how low the fourteen-foot Canadian canoe sat in the water, just a few inches of fiberglass between me and the great saurians, close relatives of the ancient dinosaurs. Not long ago, saltwater crocodiles were considered endangered, as virtually all mature animals were shot out of the rivers and lakes of Australia's north by commercial hunting. After a decade and more of protection, their numbers are beginning to burgeon. They are now the most plentiful of the large animals of Kakadu National Park, which preserved a major area of their breeding habitat. I was actively involved in the struggle to keep such places, and for me, the crocodile was a potent symbol of the power and integrity of this place and the incredible richness of its aquatic habitats.

After hours of searching the maze of shallow channels in the swamp, I was unable to locate the clear channel leading to the rock art site, as shown on the ranger's sketch map. When I pulled my canoe over in driving rain to a rock outcrop rising out of the swamp for a hasty, sodden lunch, I experienced the unfamiliar sensation of being watched. Having never been one for timidity, in philosophy or in life, I decided, rather than return defeated to my sticky caravan, to explore a clear, deep channel closer to the river I had traveled along the previous day.

The rain squalls and wind were growing more severe, and several times I had to pull my canoe over to tip the rainwater out of it. The channel soon developed steep mud banks and snags, and the going was slow. Farther on, the channel opened up, eventually petering out, blocked by a large sandy bar. I pushed the canoe toward the bank, looking around carefully before getting out in the shallows and pulling the canoe up. I would be safe from crocodiles in the canoe—I had been told—but swimming, and standing or wading at the water's edge were dangerous. Edges are one of the crocodile's favorite food-capturing places. I saw nothing, but the feeling of unease that had been with me all day intensified.

The rain eased temporarily, and I picked my way across a sandbar covered with scattered scrub to see what there was of this puzzling place. As I crested the gently sloping dune, I was shocked to glimpse the muddy brown waters of the East Alligator River gliding silently only a hundred yards ahead of me. The channel I followed had evidently been an anabranch and had led me back to the main river. Nothing stirred along the riverbank, but a great tumble of escarpment cliffs up on the other side of the river caught my attention. One especially striking rock formation—a single large rock balanced precariously on a much smaller one—held my gaze. As I looked, my whispering sense of unease turned into a shout of danger. The strange formation put me sharply in mind of two things: the indigenous Gagadgu owners of Kakadu, whose advice about coming here I had not sought, and of the precariousness of my own life, of human lives. As a solitary specimen of a major prey species of the saltwater crocodile, I was standing in one of the most dangerous places on the face of the earth.

I turned decisively to go back the way I had come, with a feeling of relief. I had not found the rock paintings, I rationalized, but it was too late to look for them now. The strange rock formation presented itself instead as the telos of the day. I had come here, I had seen something interesting, now I could go, home to caravan comfort.

As I pulled the canoe out into the main current, the torrential rain and wind started up again; the swelling stream would carry me home the quicker, I thought. I had not gone more than five or ten minutes back down the channel when, rounding a bend, I saw ahead of me in midstream what looked like a floating stick—one I did not recall passing on my way up. As the current moved me toward it, the stick appeared to develop eyes. A crocodile! It is hard to estimate size from the small nose and eye protrusions the crocodile leaves, in cryptic mode, above the waterline, but it did not look like a large one. I was close to it now but was not especially afraid; an encounter would add interest to the day.

Although I was paddling to miss the crocodile, our paths were strangely convergent. I knew it was going to be close but was totally unprepared for the great blow that came against the side of the canoe. Again it came, again and again, now from behind, shuddering the flimsy craft. I paddled furiously, but the blows continued. The unheard of was happening, the canoe was under attack, the crocodile in full pursuit! For the first time, it came to me fully that I was prey. I realized I had to get out of the canoe or risk being capsized or pulled into the deeper water of mid channel.

The bank now presented a high, steep face of slippery mud, difficult to scale. There was only one obvious avenue of escape, a paperbark tree with many low branches near the muddy bank wall. I made the split-second decision to try to leap into the lower branches and climb to safety. I steered the canoe over to the bank by the paperbark and stood up ready to jump. At the same instant, the crocodile rushed up alongside the canoe, and its beautiful, flecked golden eyes looked straight into mine. Perhaps I could bluff it, drive it away, as I had read of British tiger hunters doing. I waved my arms and shouted, "Go away!" (We're British here.) The golden eyes glinted with interest. I tensed for the jump and leapt. Before my foot even tipped the first branch, I had a blurred, incredulous vision of great toothed jaws bursting from the water. Then I was seized between the legs in a red-hot pincer grip and whirled into the suffocating wet darkness below.

The course and intensity of terminal thought patterns in near-death experiences can tell us much about our frameworks of subjectivity. A subjectively centered framework capable of sustaining action and purpose must, I think, view the world "from the inside," structured to sustain the concept of a continuing, narrative self; we remake the world in that way as our own, investing it with meaning, reconceiving it as sane, survivable, amenable to hope and resolution. The lack of fit between this subject-centered version and reality comes into play in extreme moments. In its final, frantic attempts to protect itself from the knowledge that threatens the narrative framework, the mind can instantaneously fabricate terminal doubt of extravagant, Cartesian proportions: *This is not really happening. This is a nightmare from which I will soon awake.* This desperate delusion split apart as I hit the water. In that flash, when my consciousness had to know the bitter certainty of its end, I glimpsed the world for the first time "from the outside," as a world no longer my own, an unrecognizably bleak landscape composed of raw necessity, that would go on without me, indifferent to my will and struggle, to my life or death.

Few of those who have experienced the crocodile's death roll have lived to describe it. It is, essentially, an experience beyond words of total terror, total helplessness, total certainty, experienced with undivided mind and body, of a terrible death in the swirling depths. The crocodile's breathing and heart metabolism is not suited to prolonged struggle, so the roll is an intense initial burst of power designed to overcome the surprised victim's resistance quickly. Then it is merely a question of holding the now feebly struggling prey under the water a while for an easy finish to the drowning job. The roll was a centrifuge of whirling, boiling blackness, which seemed about to tear my limbs from my body, driving water into my bursting lungs. It lasted for an eternity, beyond endurance, but when I seemed all but finished, the rolling suddenly stopped. My feet touched bottom, my head broke the surface, and spluttering, coughing, I sucked at air, amazed to find myself still alive. The crocodile still had me in its pincer grip between the legs, and the water came just up to my chest. As we rested together, I had just begun to weep for the prospects of my mangled body, when the crocodile pitched me suddenly into a second death roll.

When the tearing, whirling terror stopped again (this time perhaps it had not lasted quite so long), I surfaced again, still in the crocodile's grip, next to the stout branch of a large sandpaper fig growing in the water. I reached out and held onto the branch with all my strength, vowing to let the crocodile tear me apart rather than throw me again into that spinning, suffocating hell. For the first time I became aware of a low growling sound issuing from the crocodile's throat, as if it were angry. I braced myself against the branch ready for another roll, but after a short time the crocodile's jaws simply relaxed. I was free. With all of my power, I used my grip on the branch to pull away, dodging around the back of the fig tree to avoid the forbidding mud bank, and tried once more the only apparent avenue of escape, to climb into the paperbark tree.

As in the repetition of a nightmare, when the dreamer is stuck fast in some monstrous pattern of destruction impervious to will or endeavor, the horror of my first escape attempt was exactly repeated. As I leapt into the same branch, the crocodile again propelled itself from the water, seizing me once more, this time around the upper left thigh. I briefly felt a hot sensation before being again submerged in the terror of the third death roll. Like the others, it stopped eventually, and we came up in the same place as before, next to the sandpaper fig branch. I was growing weaker, but I could see the crocodile taking a long time to kill me this way. It seemed to be intent on tearing me apart slowly, playing with me like a huge growling cat with a torn mouse. I did not imagine that I would survive, so great seemed its anger and its power compared to mine. I prayed for a quick finish and decided to provoke it by attacking it with my free hands. Feeling back behind me along the head, which still held my body in its jaws, I encountered two lumps. Thinking I had the eye sockets, I jabbed my thumbs into them with all my might. They slid into warm, unresisting holes (which may have been the ears or perhaps the nostrils), and the crocodile did not so much as flinch. In despair, I resumed my grasp on the branch, dreading death by slow torture. Once again, after a time, I felt the crocodile jaws relax, and I pulled free.

I knew now that I must break the pattern. *Not* back into the paperbark. Up the impossible, slippery mud bank was the only way. I threw myself at it with all of my failing strength, scrabbling with my hands for a grip, failing, sliding, falling back to the bottom, to the waiting jaws of the crocodile. I tried a second time and almost made it before sliding back, braking my slide two-thirds of the way down by grabbing a tuft of grass. I hung there, exhausted, defeated. *I can't make it, I thought. It'll just have to come and get me.* It seemed a shame, somehow, after all I had been through. The grass tuft began to give way. Flailing wildly to stop myself from sliding farther, I found my fingers jamming into the soft mud, and that supported me. This was the clue I needed to survive. With the last of my strength, I climbed up the bank, pushing my fingers into the mud to hold my weight, reached the top, and stood up, incredulous. I was alive!

Escaping the crocodile was not by any means the end of my struggle to survive. I was alone, severely injured, and many miles from help. During the struggle, I was so focused on survival that the pain from the injuries had not registered. As I took my first urgent steps away from the vicinity of the crocodile, I knew something was wrong with my leg. The bastard's broken my knee. I did not wait to inspect the damage, but took off away from the crocodile in the direction of the ranger station.

After putting more distance between myself and the crocodile, I felt a bit safer and stopped to find out what was wrong with my leg. Now I was aware for the first time of how serious my wounds were. I did not remove my clothing to see the damage to the groin area inflicted by the first hold. What I could see was bad enough. The left thigh hung open, with bits of fat, tendon and muscle showing, and a sick, numb feeling suffused my entire body. I tore up some of my clothing to try and bind the wounds up and made a tourniquet for the thigh to staunch the bleeding, then staggered on, thinking only of getting back to the ranger station. Still elated from my escape, I imagined myself, spattered with blood and mud, lurching sensationally into the station. I went some distance before realizing with a sinking heart that I had crossed the swamp above the station in the canoe and that without it I could not get back to the station under my own steam. Perhaps I would die out here after all.

I would have to rely on being found by a search party, but I could maximize my chances by moving downstream toward the swamp edge, about three kilometers away. Still exhilarated by my escape, perhaps now I had a chance of survival. I had recently been reading Robert Graves' memoir of soldiers in the First World War who had been able to walk long distances with severe injuries and survived. Walking was still possible, and there was nothing better to do. Whenever I lay down to rest, the pain seemed even worse. I struggled on, through driving rain, shouting for mercy from the sky, apologizing to the angry crocodile, calling out my repentance to this place for the fault of my intrusion. I came to a flooded tributary and had to make a large upstream detour to find a place where I could cross it in my weakened state.

My considerable bush experience stood me in good stead, keeping me on course (navigating was second nature), and practiced endurance stopped me from losing heart. As I neared the swamp above the ranger station after a journey of several hours, I began to black out and had to crawl the final distance to its edge. I could do no more for myself; I selected an open spot near the swamp edge, and lay there in the gathering dusk to await what would come. I did not expect a search party until the following day, and I doubted I could possibly last the night.

The heavy rain and wind stopped with the onset of darkness, and it grew perfectly still. Dingos howled, and clouds of mosquitoes whined around my body. I hoped to pass out soon, but consciousness persisted. There were loud swirling noises in the water, and I knew I was easy meat for another crocodile. After what seemed like a long time, I heard the distant sound of a motor and saw a light moving across the other side of the swamp. Thinking it was a boat crossing the swamp to rescue me, I had enough strength to rise up on my elbow and call out for help. I thought I heard a very faint reply, but then the motor grew fainter and the lights went away. Thinking I had imagined the voice, I was as devastated as any castaway who signals desperately to a passing ship and is not seen.

It was not from a boat that the lights had come. Passing my caravan, the ranger noticed there was no light. He had come down to the canoe launch site in a motorized trike to check and, realizing that I had not returned, stopped his motor to listen. He had heard my faint call for help across the dark water, and after some time, a rescue craft appeared. As they lifted me into the boat that was to begin my thirteen-hour journey to Darwin Hospital, my rescuers discussed the need to go upriver the next day and shoot a crocodile. I spoke strongly against this plan: I was the intruder on crocodile territory, and no good purpose could be served by random revenge. The area was full of crocodiles: Rangers had reported sighting, in the boat searchlight, a number of large crocodiles in the water just around the spot where I was lying. That spot was under six feet of water the next morning, flooded by the rains signaling the onset of the wet season.

In the end I was found in time and survived against many odds, thanks to the ranger's diligence, my own perseverance, and great good fortune. A similar combination of good fortune and human care enabled me to overcome an infection in the leg that threatened amputation or worse. I probably have Paddy Pallin's incredibly tough walking shorts to thank for the fact that the groin injuries were not as severe as the leg injuries. I am very lucky that I can still walk well and have lost few of my previous capacities. Lady Luck shows here, as usual, her inscrutable face: Was I lucky to survive or unlucky to have been attacked in the first place? The wonder of being alive after being held—quite literally—in the jaws of death has never entirely left me. For the first year, the experience of existence as an unexpected blessing cast a golden glow over my life, despite the injuries and the pain. The glow has slowly faded, but some of the gratitude for life it left will always be there, even if I remain unsure whom I should thank. The gift of gratitude came from that searing flash of near-death knowledge, a glimpse "from the outside" of the alien, incomprehensible world in which the narrative of self has ended.

There remain many mysteries about the reasons for the attack on the canoe itself, which are unusual in crocodile lore. One issue on which there has been much speculation is the size of the crocodile. It has always been difficult for me to estimate its size because for most of the attack, it was either only partly visible or had ahold of me from behind. The press estimate of fourteen feet—which they arrived at somehow and published widely some five days before I made any sort of statement on the subject—was, I think, certainly an overestimate. One glimpse of the partly submerged crocodile next to the fourteen-foot canoe suggested that it was not as long as the canoe. If the press had an interest in exaggerating the size of the crocodile, the park service, which feared legal liability, had an interest in minimizing it; neither group seemed interested in my views on the matter.

The park service speculated that the crocodile may have been a young male evicted from breeding territory and perhaps embattled by other crocodiles. Their theory is that the crocodile attacked my canoe after a collision by mistaking it for one of these older aggressive crocodiles. From my perspective, however, there are some problems with this account. It is very unlikely that I accidentally struck it with the canoe, as the story assumes and as some press reports claimed. Crocodiles are masters of water, and this one was expecting me and saw me coming. The crocodile most likely observed the passage of my canoe on the way up the channel only a short time before, as it did seem to *intercept* the canoe. Why should a small crocodile of less than ten feet aggressively attack a much larger, fourteen-foot canoe "crocodile" unless we assume that it was bent on suicide? The smaller the crocodile, the more implausible such an attack story becomes. Because crocodiles become sexually mature at around ten feet, the park service's minimization story of a "self-defense" attack by a "small" evicted crocodile is not even internally consistent. My personal estimate is that it was probably a medium crocodile in the range of eight to twelve feet.

Possible explanations for the anomalous attack are almost limitless. Perhaps the crocodile's motives were political, against a species-enemy; human beings are a threat to crocodiles, of a more dire kind than crocodiles are to human beings, through the elimination of habitat. The crocodile may have thought that any human being who ventured alone into these waters in those conditions was offering itself as a sacrifice to crocodile kind. The extreme weather events may have played a role. The crocodile is an exploiter of the great planetary dualism of land and water. As Papua New Guinea writer Vincent Eri suggests in his novel, *The Crocodile,* the creature is a sort of magician: Its technique is to steal the Other, the creature of the land, away into its own world of water where it has complete mastery over it. Water is the key to the crocodile's power, and even large crocodiles rarely attack in its absence.

The crocodile is then a boundary inhabitant and may take a person in a canoe as either of the land or of the water. If a crocodile perceives such a person as *outside* its medium of mastery, the person may not be seen as prey and may be safe from attack. If a person in a canoe is perceived as potentially of the water, as he or she might easily be in an early wet-season day of torrential rain when the boundaries of the crocodile empire are exploding, the person may be much less than safe. Clearly, we must question the assumption, common up until the time of my attack, that canoes are as safe as larger craft because they are perceived similarly as outside the crocodile's medium.

The most puzzling question of all, of course, is why the crocodile let me go. I think there are several factors here. Because it was not a large crocodile that can kill with little effort, perhaps I was marginal prey. The depth of the water and the way it had ahold of my body made it hard for it to keep me under, and it may have let me go the first time to try to get a better grip higher up on my body. Its failure to keep me submerged suggests that it could have underestimated my size, seeing me sitting in the canoe, or overestimated the depth of the rapidly rising water. Maybe it was a stray or a newcomer to the area who did not know the terrain well and was not familiar with the good drowning spots in the shallower back channels. My friend the sandpaper fig allowed me to retain a determined grip on my own medium and contributed essentially to my survival. In another encounter in the territory a few years earlier, an adult man was saved from a fourteen-foot crocodile dragging him off in shallow water by the grasp of a ten-year-old girl pulling the opposite way.

Perhaps, too, the crocodile let go its hold because it was tiring; I experienced the crocodile through the roll as immensely powerful, but that intense burst of energy cannot be sustained long, and must accomplish its purpose of drowning fairly quickly. I have no doubt that had the crocodile been able to keep me submerged after the first roll, there would have been no need for a second. My advice for others similarly placed is the same as that of Vincent Eri, who uses the crocodile as a metaphor for the relationship between colonized indigenous culture and colonizing Western culture. If the crocodile-magician-colonizer can drag you completely into its medium, you have little chance; if you can somehow manage to retain a hold on your own medium, you may survive.

I had survived the crocodile attack but still had to survive contest with the cultural drive to represent such experiences in terms of the masculinist monster myth: the master narrative. The encounter did not immediately present itself in the guise of a mythic struggle: I recall thinking with relief, as I struggled away from the attack site, that I would now have a good excuse for being late with an overdue article and would have a foolish but unusual story for my small circle of walking companions. Nor did the crocodile appear as an implacable monster, although its anger was a mystery. I figured I might have offended it somehow and wished we could have communicated. It had, after all, eventually let me go. Because crocodile attacks, especially in North Queensland, have often been followed by episodes of massive crocodile slaughter in which entire river populations were wiped out, I feared similar reprisals and felt not victorious, but responsible for putting the crocodiles at risk. To minimize this risk, I tried hard at first to minimize media publicity and keep the story for my friends' ears alone.

This proved to be extremely difficult. The media machine headlined a garbled version of the encounter anyway, and I came under great pressure, especially from the hospital authorities whose phone lines had been jammed for days, to give a press interview. We all want to pass on our story, of course, and I was no exception. Often, for the dying, it is not death itself that is the main concern, but the loss of their story, the waste of the narrative that is their life's experience, the crucial ingredients it might contribute to the salty stock of human wisdom. During those incredible split seconds when the crocodile dragged me a second time from tree to water, I had a powerful vision of friends discussing my death with grief and puzzlement. The focus of my own regret was that they would never learn how my story had ended. They would never know about my struggle and might think I had been taken while risking a swim. So important is the story and so deep the connection to others, carried through the narrative self, that it haunts even our final desperate moments.

To the extent that the story is crucial, by the same token the narrative self is threatened with invasion and loss of integrity when the story of the self is taken over by others and given an alien meaning. This is what the mass media tend to do in stereotyping and sensationalizing stories like mine, and this is what is done all the time to subordinated groups, such as indigenous peoples, when their voices and stories are digested and repackaged in assimilated form. As a story that evoked the monster myth, mine was especially subject to masculinist appropriation. The imposition of the master narrative appeared in a number of different forms: in the exaggeration of the crocodile's size, in the portrayal of the encounter as a heroic wrestling match, and especially in its sexualization. The events seemed to provide irresistible material for the pornographic imagination, which encouraged male identification with the crocodile and interpretation of the attack as sadistic rape. The reinterpretation of the experience in these sexual terms and its portrayal in porno films like *Crocodile Blondee* reveal the extent to which sadism is normalized in dominant culture as masculinist sexuality.

Although I had survived in part because of my active struggle and bush experience, one of the major meanings evoked by the narrative was that the bush was no place for a woman. Much of the Australian media seemed to have trouble coming to terms with the idea of women being competent in the bush, but the most advanced expression of this masculinist mindset was *Crocodile Dundee,* which was filmed in Kakadu not long after my encounter. If page-three articles and *Crocodile Blondee* eroticized the crocodile as a male sadist, *Crocodile Dundee* took the more respectable course of eroticizing female passivity and victimhood. The two biggest recent escape stories had both involved active women, one of whom had actually saved a man. However, the film's story line split the experience along conventional gender lines, appropriating the active struggle, escape, and survival parts of the experience for the male hero and representing the passive "victim" parts in the character of an irrational and helpless woman who is incompetent in the bush and has to be rescued from the crocodile-sadist (the rival male) by the bushman hero.

For a long time, I felt alienated from my own story by the imposition of these stereotypes and had to wait nearly a decade before I felt able to repossess my story fully and write about the experience in my own terms. For our narrative selves, being able to pass on our stories in an authentic form is a crucial part of satisfaction in life, a way to participate in and be empowered by culture. Passing on the story is often much more important to people than material possessions; that is why they are capable of making such enormous sacrifices so that the story, the struggle of memory against forgetting, will survive. Retelling the story of a traumatic event can have tremendous healing power. During my recovery, it seemed as if each telling took part of the pain and distress of the memory away. Passing on the story can be a way to transcend not only social harm, but also our own biological death and bodily limitation.

Cultures differ considerably in the opportunities they provide for passing on their stories. Because of its highly privatized conception of the individual, contemporary Western culture is, I think, relatively impoverished in this respect, especially compared with certain indigenous cultures. That is part of the emptiness at its core, an emptiness that must be assuaged with more material commodities and more control. In contrast, many Australian aboriginal cultures seem to offer rich opportunities for passing on the story, in ways that are reflected in the opportunities for the transcendence of individual death provided in their accounts of human identity. In much Aboriginal thinking about death, animals, plants, and humans are seen as sharing a common life force, and there are many narrative continuities and interchanges between humans

and other life. To the extent that such a culture provides intense narrative identification with land and with an ecological and social community based on the
history of that land extending over time, a form of individual meaning and narrative continuity is available that enables some degree of transcendence of the
individual's death.

In Western thinking, in contrast, the human is set apart from nature as radically other. Religions like Christianity must then seek narrative continuity for
the individual in the idea of an authentic self that is nonbodily and above the
earth: the eternal soul. This sort of recipe for transcendence of death, however,
is bought at a great price, one that can provide narrative continuity for the individual only in isolation from the cultural and ecological community and in
opposition to a person's perishable body. This solution to the problem of providing continuity depends on creating a split within the individual and within
culture between the immaterial, eternal soul, as the "higher" human essence
that continues after death, and the inessential, devalued, and animal body,
which dies. Boundary breakdown is an ever-present threat and source of anxiety, reflected in the aura of horror that surrounds deathly decay in the Western
tradition, as the forbidden mixing of these hyperseparated categories, the disso-
lution of the sacred-human into the profane-natural.

If ordinary death is a horror, death in the jaws of a crocodile is the ultimate
horror. It multiplies these forbidden boundary breakdowns, combining decomposition of the victim's body with the overturning of the victory over nature
and materiality that Christian death represents. Crocodile predation on humans threatens the dualistic vision of human mastery of the planet in which
we are predators but can never ourselves be prey. We may daily consume
other animals in their billions, but we ourselves cannot be food for worms and
certainly not meat for crocodiles.

For the human self as narrative subject, the end of the story is beyond com-
prehension, outside the framework. For me, it was as if, before the event, I
saw the whole universe as framed by my own narrative, as though the two
were joined perfectly and seamlessly together. As my own narrative and the
larger story in which it was embedded were ripped painfully apart, I glimpsed
beyond my own realm a shockingly indifferent world of necessity in which I
had no more significance than any other edible being. The thought, *This can't
be happening to me. I'm a human being, not meat, I don't deserve this fate!* was one
component of my terminal incredulity. Confronting the brute fact of being

prey, together with the astonishing view of this larger story in which my "normal" ethical terms of struggle seemed absent or meaningless, brought home to me rather sharply that we inhabit not only an ethical order, but also something not reducible to it, an ecological order. We live by illusion if we believe we can shape our lives, or those of the other beings with whom we share the ecosystem, in the terms of the ethical and cultural sphere alone.

Although I had been a vegetarian for some ten years before the encounter with the crocodile and remain one today, this knowledge makes me wary of the kind of uncontextualized foundation for vegetarianism that suggests that predation is either a negligible anomaly or an unredeemable ethical deficiency in the ecosystem. The presentation of the food chain as a (potentially) peaceful order ideally subject to nonviolent reconfiguration leads ultimately toward the thoroughly antiecological position that the earth is ethically improved by the elimination of predation. Such an overgeneralized form of vegetarianism can remain consistent only by redoubling the stress on certain radical discontinuities in the categories of plants, animals, and humans. It must situate humans as exclusive possessors of an ideal ethical nature denied to other "lower" plant and animal forms of life and, on pain of starvation, reemphasize the Cartesian consciousness boundary to include in the field of ethical treatment only sentient beings. Other living beings remain outside the sphere of ethical eating and, presumably, ethics. These moves leave largely unquestioned, or merely relocate, the radical discontinuities of the dominant culture, in which the truly human belongs to an ethical order beyond edibility and ecology, while the nonhuman belongs to a hyperseparated edible and ecological order unconstrained by ethics.

In accordance with this dominant view, we have split both ourselves and the world into hyperseparated realms of nature and culture and have acted as if only the last were genuinely human and truly important. To realize a truly human ecological identity, we must overcome this radical split, knitting together these hyperseparated realms from both directions, both by extending ecology to the human sphere and by extending ethics to the nonhuman sphere. This means that we must acknowledge not only that ethics applies to those we eat (both plant and animal life), but also the possibility of forms of ethical reciprocity in the food chain. In this case, ethical eating may not always exclude the taking of life, and predation may take forms that are understood in ethical terms. The project of constructing an ethical form for an ecologically sensitive life requires of us ethical evaluation of our ecological identities and relationships but does not suggest that the world would be ethically improved by the elimination of predation. I am a vegetarian primarily because ethical and ecological forms of predation are only exceptionally available in contemporary Western society, with its factory farming and commodified relationships to food.

Resisting the identification of the human with the ethical, and the nonhuman with the ecological, opens a way to bring the spheres of the ethical and ecological, culture and nature, closer together, but it does not make them identical. Paradox envelops their meeting; in that paradox we must somehow make our home as beings of both nature and culture whose ethical life should begin with the recognition that "all our food is souls." Coming to terms with the ethical challenge of other large predators is part of coming to terms with this paradox and with the ethical dimensions of our own predation. In the large predator of humans, the ethical and the ecological collide; we are forced to face an ecological challenge to the realm of ethics and to try to respond with something more ethical than condemnation or revenge. This is part of the mystery and fascination of our relationship with other large predators.

Large predators like lions and crocodiles are the subjects of an obsessive gaze in contemporary popular culture, and some of this may be derived from the contemporary political agenda of social Darwinism. However, there are also some good reasons for the focus on large predators: They present an important test for us, in both ethical and ecological terms. As ecologists have stressed, the ability of an ecosystem to support large predators is a criterion of its ecological integrity. Crocodiles and other creatures that can take human life also present an important test of ethical and political integrity. The colonizer identity is positioned as an eater of Others who can never themselves be eaten, just as the unmarked gaze of the colonizer claims the power to see but not to be seen. In terms of virtue ethics, the existence of free communities of animals that can prey on humans indicates our preparedness to share and to coexist with the otherness of the earth, to reject the colonizer identity and the stance of assimilation, which aims to make the Other over into a form that eliminates all friction, challenge, or consequence. The persistence of predator populations tests our integration of ethical and ecological identities, our recognition of our human existence in mutual, ecological terms, as ourselves part of the food chain, eaten as well as eater.

Thus the story of the crocodile encounter has, for me, come to have a significance quite the opposite of what is conveyed in the master/monster narrative. For me, it is a cautionary tale about survival and our relationship with the earth, about the need to learn to recognize who we are in different terms that acknowledge our own animality and the ecological as well as ethical context of our lives. I learned many personal lessons from the event, one of which is to know better when to turn back and to be more open to the sorts of messages and warnings I had ignored on that particular day. As on the day itself, so even more to me now a decade later, the *telos* of these events lies in the strange rock formation, which symbolized so well the lessons about the vulnerability of humankind I had to learn, lessons it seems largely lost to the technological culture that now dominates the earth. In my work as a philosopher, I see more and more reason to stress our failure to perceive this vulnerability and the distortions of our view of ourselves as rational masters of a tamed and malleable nature. The balanced rock suggests a link between my insensitivity to my own vulnerability and the similar failure of my culture in its occupation of the planetary biosystem. Let us hope that it does not take a similar near-death experience to instruct our culture in the wisdom of the balanced rock.

Doug DuBois

Tuna Country

Charles Bowden

Photographs by Doug DuBois

I have never known fear, but she and I spend a lot of time together. She's more of doer than a talker, and so it's damn hard to make notes what with the whirligig she creates when the big frights are on her. So I can't say I really know her. But I'll tell you a story. On July 20th at 4:30 A.M., fifteen cops came and took a woman. Thank God it was summer, since she wasn't wearing anything but a bra and panties. Later, folks washed the blood off the walls. The cops were square guys; they left her eight-week-old baby. They found the woman ten days later in a fifty-gallon drum filled with acid. Her driver's license was of no help, but the registration numbers on her breast implants identified her. These details are the kind of thing that gets in the way of the big picture, and we all know that mackerel skies are ending and the macro world is the coming idea. The thing to remember is that barrel. Two kids were walking along the sewage canal and saw it. They fished it out thinking they could sell the barrel for a few bucks. And here you thought the new earth was a metaphor.

As I walk down the corridor, the lights licking at my face and body are green, red, and yellow. My hand has been stamped with indelible ink; I have paid my seven pesos. At the end of the corridor is a cavern of thunder with maybe a thousand boys and girls standing like statues under the strobes of the dance hall. The band plays disco, rancheras, rock and roll, all manner of beats to make the body move and the skin go hot and moist. I am in a chamber of lust, not simply love, not pleasure, but something deeper, more driven, more moral. Lust. Many of us now fear this thing called lust, but then many of us have not yet earned a ticket to this chamber. When we do, we will also lust. You see, there are forces so powerful they can push us past the defenses of diet and nutrition and take us into another country. Do not worry, you will know when it happens.

This place is a few blocks from the official border that separates the United States and Mexico and is tucked away in the whoring district of Juárez, the Zona Rosa. I am here as a historian. For some weeks or months, young girls have disappeared from this place and then were found dead and in sorry condition. But no one here is keen on history. They are factory workers, teenagers in the American-owned maquiladoras *of this city. They make three or four or five dollars a day . . . fifteen dollars a day if you ask a government or a trade association or anyone except the workers themselves. They work six days a week, and at night they feed their dreams. They come here with young bodies, see-through blouses, tight Levi's, boots, cowboy hats, shorts and halters, satiny red dresses, high heels, perfume, rodeo belts, stitched and elaborate* vaquero *shirts, a few pesos. And dreams. Everything but history and a floor of understanding. Still, they know the names of men on the monuments scattered around town, they know their families and villages, they know the United States is very rich, they know they are poor, and they know they work six days a week and they need to feel love or believe in love or fuck or something like love. And they know the name of this place where they drink beer and dance.*

Tuna Country. That's right. Tuna Country.

We fear Tuna Country. We say to ourselves, yes, it exists, it is over there, a by-product of the orderly development of modern industrial society. Or it is a perversion soon to pass into nothingness. Or it is a wrong, and we will study it or outlaw it or regulate it or kill it. The girl, and by God she is a girl of fourteen or fifteen, standing there in the shiny new dress with a slit up the side to her hip and her ass thrust out when she moves on those high heels, that girl looking into our eyes with hope and lust and mercenary dreams, she actually exists and she is not to be denied by our schooling. We have tried to confine her with our zoning for red-light districts, with our police and military boundaries of national borders, and she pisses on all of us and breaks our limits and streaks into the country of desire.

She wants, and she will have what she wants or die trying. That is the musk floating in the smoky air of Tuna Country. In a time of definitions that are palpable frauds—the natural world, the industrial world, the left, the right, the information society, the First World, the Third World, and all the other goddamn worlds—Tuna Country is an affront and a release. It assaults our tidy concepts with something we all know—hunger.

The cop stops us on the sidewalk outside Tuna Country and expresses deep concern over our open cans of beer. He has the gun and the badge and means there are rules, natural law, gods, straight-and-narrow ways, and always force. We apologize, say we mean no harm, and he passes like a plague floating over the city. About two hundred yards behind my back, there, right across the plaza in front of the cathedral, that is where a girl disappeared at three in the afternoon on a sunny day and vanished into the boneyard. Just across the narrow street where I insult the law with my open beer, there in that cantina is where the men drank and laughed and are said to have planned the kidnapping, rapes, and murders of some of the local muchachas. *I have a photo: The cop stands in the foreground holding a high-heeled shoe. He wears sun glasses and looks professional. Behind him other officers gather around the sprawled body in the desert. So I know there is order and realize the transgression of my open beer at 3* A.M.

The nights frighten some, and the days seem to help not at all. The heat has come early, and the desert bakes in April and bakes some more in May and continues to writhe under a cloudless sky. But she does not speak of such obvious facts. She is telling me of her garden in the village where she lives by a *cienaga* five or six scant miles north of the border. Of course, the grass has long been devoured, the water holes are empty sockets of cracked earth, and herds have been sold or gone to bone. In the Mexican north during the parched times of 1995, at least three hundred thousand cattle looked at the sun, slowly moved their swollen tongues, and fell dead from thirst. This recurring music has for centuries been a somber score underlying the local dance of life. That is what leads to talk of her garden, a normally fabulous place with armies of chiles, tomatoes, corn, beans, and squash. She came into this country more than a quarter century ago, riding a rainbow, and lived in a tepee and various shacks before finally coming to ground by the bog. She has raised up dynasties of gardens. But this year, the hunger is too great, she notes in her brisk and clipped speech.

Wings flapping over Puerto Rico, obscene outlines of lust against the blood red sky of dusk, the beasts lift and swirl and then with a roar head toward the mainland. The

Mexicans begin to note their dead, bodies sucked dry of blood, throats torn. The night offers new danger. They call them chupacabras, *and they believe in them because we must all believe in something. The televisions shout* chupacabras, *the radios, the newspapers all agree. God, they have come at last to save us, these blood-sucking monsters that stir in the night and give us the hope that life matters, that life is mysterious, and that we are not in control or responsible for what happens in the night or what we do with our hours. Soon the* chupacabras *are sighted on the American border. Surely, they will storm north saving the gringos from the emptiness, sucking blood state-by-state, and then downing the moose, the elk, the musk-ox of Canada. As the money dies and the jobs do not pay, and the governments talk to themselves or to no one at all, the* chupacabras *bring the gift of life with meaningless moments of death. We are not in control. See?*

Every night without fail the deer and the javelina storm my friend's defenses and pillage her cultivated ground near that international fantasy line. There is nothing to be done since there is so little left out on the land. She fears the rains will skip this summer, no cool clouds will open up and pour down on the land. The sense of a deep burning is everywhere. The forests have been ablaze since April, and the merest spark will send walls of flame marching through what is left of the grass. Clearly, there is a pattern. Things are gathering in as the land itself shudders and dries and teeters on the edge of some vast mummification. This is to be expected of course—there is evidence in the tree-ring histories of droughts lasting fifty years, a hundred years, 250 years. Vast and harsh reckonings have come to this place before. The deer and wild pigs cannot be stopped, she says. The coyotes grow yet more bold and desperate.

She is reconciled to these matters. Who can lift their hand and stay this dry and ferocious storm? And perhaps it is all just as well, this reckoning. Perhaps it is time to stop the lies and feel hard truths rake across the flesh.

I live to the whir of hummingbird wings, and my nights are choked by the scent of blooming cactus. The sun roasts my skin, and I never block its hunger. In winter, the floors go cold, and stones bark at my naked skin. I have never known the natural world, just the pleasures of the flesh, and so I am ill-prepared for this era of categories. I am not cut out for peace or ecology or conservation or harmony or safe places that wall off a throbbing world. What I can do is lust, lust for love, lust for justice, lust for music, lust for strong drink, hot food, and star-filled nights. I can lust for anything except escape. I do not believe in droughts or floods. I accept all manner of weather. All of it, absolutely all of it. But I am partial to the storm. So I wander easily in the hills, do not trust the coyotes or wish them ill. And I fish in Tuna Country on a catch-and-release basis. One more thing: I am not good at boundaries. And good fences don't make good neighbors; they don't make neighbors at all.

On February 23 in Ciudad Juárez, the body of a twenty-five-year-old man is found with hands tied behind his back, his eyes taped, his torso tortured. The day before the local police and the Mexican federal police had fought a battle over who had the right to shake down the district's whores, drug dealers, merchants, and other prey. Another man about thirty was found with six bullets in his head. Some think they were police informers who failed at their work.

They offer me two bulging file drawers full of their reports, analyses, studies, projections, seminar workups, and official proclamations. I have served my time in the analog world. Worst drought in God knows how many years, reservoirs empty, agribusinesses in Texas and Oklahoma going down to dull-witted death like so many dinosaurs befuddled by change. Rivers disappearing and leaving sandy wakes that writhe like the tracks of giant snakes. There is talk of greenhouse effects, permanent weather shifts, seas rising to drown the coastal Babylons, vast ice sheets plunging into oceans of Antarctica, a vicious sun raising cancers on innocent flesh.

We will gather by the river, the beautiful, beautiful river . . .

I am electric with the dry. I move for weeks through a region of hazy skies as the mountains burn from tinder loads fostered by almost a century of human fire suppression. I have been waiting, oh so long. Waiting for a judgment day. The deer and javelina storming the garden, the coyotes grown so bold, all are my sign. We cannot do what we have done here without a payback. We have planted an urban civilization in a very dry place. We have fostered an agricultural empire in a region of deep droughts periodically broken up by moments of rain. We have unleashed too many cows on too much dry land. We have pissed in the face of reality. Judgment, there is sure to be one, God or no God. In early April, the dry and hungry souls of Ciudad Juárez on the line at El Paso imported a Cherokee medicine man from Oklahoma. He danced, he sang. The rains still did not come. What came instead were creatures that flew in the night and sucked the blood clean out of animals. Monsters boiled out of the earth. The creatures have been seen, their dead found in the light of dawn. True, no one has produced a single specimen of this new creature. But this does not bother anyone, and their existence is widely believed all across the Mexican north. The *chupacabras,* goat suckers. Because these vampire monsters make sense. The year before, the panic sweeping the Mexican north was that children were being snatched from school yards so that they could be slaughtered and their body parts sold to rich gringos who needed organ replacements because of their vile and filthy habits. Over the decades, I have seen these phantoms come in many forms. For a while in the mid-eighties there were rumors of American criminals stealing children from the Mexican north so they could drain their pure and innocent blood to sell to AIDS sufferers in the United States. There are patterns here, but no one likes to look at patterns. Still, the feeling of dread among the many is palpable. The very air confirms these sensations. Just look out the window into the blaze of white light.

The North Americans have different monsters. They devour narcotics and denounce pushers. There are many others, also. We have wetbacks, or illegal aliens, or undocumented workers. We are so frightened we constantly make up new names for them. According to reports, they do not suck the blood out of our bodies, but rather suck the resources out of our government and our society. For this reason, their children must be tossed out of school and kept ignorant, and they must be denied medical care should they fall ill. These wetbacksillegalaliensundocumentedworkers are produced by five hundred years of Spanish and Mexican corruption and by almost a century of North

American tinkering with and distortion of the Mexican economy. They are the by-product of Mexico's long effort to be an industrial nation. They are the human garbage that an industrial nation does not need, the poor people who cannot adequately consume and are not needed for the production of things to consume. And so they come, and so they are cursed, and so they become our monsters. A brown horde knocking at the midnight door. The North Americans believe these monsters are poisoning their youth with chemicals. The North Americans believe these monsters are going to destroy the English language, gerund by gerund. Machetes will hack and slash the dangling participles.

Truly, we all live in a gothic place.

An employee of the Cartel de Juárez known as El Gory was found shot and tortured the day after April Fool's Day. It is thought this is the work of the chief underboss of the Cartel de Juárez. The big boss himself is out of town. He is called El Señor de los Cielos, the boss of the skies. He earns around $200 million a week and is a very private man.

Let us pretend facts exist. Under the previous Mexican president, Carlos Salinas de Gortari, the average income declined 37 percent. Then came President Ernesto Zedillo, and things got truly hard. In the United States, per capita income in constant dollars has been on the skids since 1973. The economies on both sides of the line steadily sink, and each new depth is called a readjustment, and each new firing is called downsizing or right-sizing or nothing at all. We are several years into what *Homo sapiens* call a bad drought and what the mountains and flora and fauna of the line call weather. A vast, silent, and ignored folk movement is under way in Mexico. Millions of people are walking away from the dying ground and trudging north. They are fleeing ruin and death, but they are going toward nothing. The same thing is occurring in the American Southwest as people pour into low-wage jobs, sunshine, and trailer courts. There is no longer anything to go toward. And of course, drugs, we must face the facts of drugs. Without the cash flow from drugs, Mexico's official economy—the one monitored by those presidents and finance ministers and economists and bankers—would collapse. Thirty billion dollars of hard currency is flowing into Mexico each year from the sale of dope—a sum larger than the proceeds of all legal exports combined. There are the killings also; after all, capitalism is acknowledged by even its most rabid defenders to be an act of creative destruction. Yes, the killings are very good and commonplace. A $20 million load of dope is seized by the American authorities in Nogales, Arizona. Within days, the bodies of six executed souls show up in Nogales, Sonora. There is constant accounting, a bullet-driven reckoning. But we will get to the killings later. They are really no different than the deer storming the garden. Or the mountains afire, crops burning in the fields, the trudging of millions north. There are patterns, just take a look.

The owner of the restaurant near the federal Indian arts museum is cut down in a drive-by shooting at his place of business. There is talk the dead man was linked to a drug operation. It is the sixteenth day of April, and the sound of the six bullets that tore through his body can hardly be heard in the roar of Juárez. The splashing of his blood as he died is also soon forgotten. These things happen, and life goes on.

A week after that I am in Ciudad Juárez in a *colonia* of cardboard hovels built on a dune right by the line. There is no water, electricity is pirated from poles, and extension cords snake across the sands to the shacks. I've dropped in because five or six dead girls have shown up in the neighborhood of late—raped, strangled, and left with one shoe on and panties down around the ankles. Their hands were tied with the lace of the other shoe. Yet another psychopathic workman toiling in the dying days of this century and millennium. The American authorities are erecting a huge steel wall fifty yards away to discourage the local residents from fleeing into the United States. The *colonia* has responded by erecting a metal sign that reads, "The ten thousand residents of the pueblo of Anapra protest against this Berlin Wall. We solicit the governments of the United States and of Mexico to make an international crossing at this place." This policy statement is followed by something closer to a cry from the heart:

> I had a dream
> I saw people holding hands together
> with no iron walls but
> bridges of freedom.

The wall going up looks to be twelve feet tall. Within fifty feet of the line runs the tracks of the American railroad. As I stand in the dunes, an Amtrak train rolls by, complete with observation cars and a club car. The passengers all stare out the window at the Mexicans with their cardboard shacks and their talk of bridges of freedom. Now the people on the train have done Mexico, and they roll away toward the fleshpots of the Pacific Coast. I am at peace in Anapra. This is the real great kiva, and here I can feel the new age truly beginning. No one is chanting, but *corridos* bark from the boom boxes in the scattered shacks. In the past thirty-six months, the dreamy residents of Anapra have robbed that train six hundred times.

Julian is a friend of mine, and Tuna Country is old country to him. For years he worked in a maquiladora *and each Friday or Saturday he came to such places as this and drank and dreamed and held women in his arms. Sometimes they would fuck him. He was happy, very happy, he says. You have to understand, he goes on, the* maquila *workers are happy, they are happy fools. They know nothing but the factories, the poor wages, the sexual intrigues that crackle inside the plants, the moves by the managers to bag a free piece of ass. And the music through which they escape from themselves, transcend themselves, and in their dreams finally become themselves. A Third World economy class Saturday night fever. Two women have come to the table, one twenty-seven, the other brushing thirty. We have paid the waiter ten pesos for their company. They are not full-blown pros; they are hungry for dreams. They wear skintight miniskirts, blouses to accentuate their breasts, and the passive faces of goddesses. The younger one, Isela, has two years in a factory, a five-year-old boy, and a deep distrust of men because "they all tell lies." Now she says she is afraid because of the missing girls and the things that happened to them. She says this is the first time she has come to Tuna Country, but this is obviously not true. The waiter* *treats her as a regular. Her lie is normal, and now more than normal. No one wants to be involved in the investigation surrounding the missing girls, including my friend Julian. No one in this place really wants to know. They prefer to dream. Ah, but the dreams can be expensive. Mas cervezas! The waiter brings a bunch in the blackness of the room. A fumbling over the proper payment, a roll of bills, a few peeled off, some discussion, more bills. Settled. The next day we will discover the waiter in making change has clipped us of $200 American. We are all happy fools in Tuna Country.*

Another friend of mine gathers information and compiles it, and this information festers in him and keeps him alive. For years he tracked the murders in El Paso and Ciudad Juárez, and sometimes he can sell his work to the Mexican press. But usually no one cares to publish his research. At other times, they care very much that his research not be published. He lives in the great silence that is the dominant feature of the border. Here factories boom, but no one seems to report the wages; here killings go on, but no one admits to hearing the bark of the Uzis. We're friends. We laugh a lot, keep compiling these lists that go . . . well, you've already tasted some of this particular list.

The eighteen-year-old is tortured over the entire body—we will skip silently over the tools used and the techniques applied so skillfully—and the corpse is found on the fifth day of May in Ciudad Juárez. . . .

"This is not right," he tells me. I look at the wall behind him where he has hung a portrait of Emiliano Zapata, one in which the revolutionary wears the sash of the president of Mexico, a doodad that never graced his small, dark Indian body during his years on this earth. I'm drinking a glass of wine, and my eyes are locked on the day-to-day record he has amassed of the last six months' carnage, a record largely unreported by the newspapers, denied by the police agencies, and of no interest to anyone save him and me. I have been lost at the border for a long time and grown adjusted to its silence, violence, and nonexistence. This border craziness cannot be cured by any change in location or habits. Two weeks or so ago, I was in Salt Lake City drinking in one of those private clubs where Mormon piety has driven those with serious thirsts. Even words like *wine* and *beer* cannot be displayed on signs in the city of saints, and so merchants of booze have been driven to announcements like "WE HAVE BEERNUTS" or, my very, very favorite, "ICE COLD DEER." This time as I drink in the private club eight or nine hundred miles from the line, a defense attorney explains to me that any Mexican driving up the interstate in a new car or truck is "toast" and will be instantly pulled over by the cops and tossed about in their search for drugs. A week earlier in Tucson, a Mexican-American cop asked me over drinks, "What do you call a Mexican in a new four-wheel-drive vehicle?" He then paused and said, "Grand theft auto." There is no cure for my condition, and I seek no therapy. I see a rain of knives, a flood of blood, a seizure that rents buildings, shatters dams, swallows schools

into giant cracks in the earth. The mountains wheeze, and gusts of dust roar out of the canyons at one hundred miles an hour. The children, naturally, will be sacrificed. The women bartered. The dogs eaten. The nights soft with the sound of slicing, the purr of stilettos probing flesh like lovers. Of course, the cats will be harder and the eating not so good. More pepper, I say, and days of marinating.

This is not right, my Mexican friend always tells me. No, it is not right. But truly it is. I look at the list of killings in my hand and sip more wine. The deer are pillaging the gardens, the pigs have gone wild, and let's not think about the gleam in the eyes of the coyotes. The rains do not come. The mountains are on fire. Reckoning. Old Testament time. Gonna be the fire next time, you know. Ain't gonna be about NAFTA, ain't gonna have anything to do with border commissions, government programs, interdisciplinary projects, white power, black power, brown power, green power. Ain't gonna have truck with none of that. Past talk. Into doing, not studying. New age is upon us and is being written by illiterates with a sure hand. No one is in control now; no one can stop the fire on the mountain. Been to the mountain, and it's full of fire fighters. Been to the valley of death, and it's full of American factories and Mexican hands. I have a dream, but I've gotta take my cutting torch to the Berlin Wall right now.

On May 12, Jorge Guitierrez Román is slaughtered in the center of Juárez in his twenty-sixth year. He was a police informer and was believed murdered by a gang in the area that operates without any restriction.

We are born to believe, and as believers we survive by feeding on lies. I can sit in Juárez and drink for hours with friends and argue and laugh, and they will explain to me that everything is fucked, that the government is crooked, the police are thieves, and so forth. And then out of thin air they will suddenly start believing. Eight men will be arrested for bagging girls in Tuna Country and then taking them to the sands of Anapra and raping and killing them, and my friends will grow excited over the solution of the case and of this statement of order and logic. I will ask them, why do you believe these charges? Do you really think eight guys have been responsible for the 150 or 200 girls that have officially disappeared in the last twelve months? Do you really think these guys have confessed or simply been tortured into consent? My friends will dimly nod and finally agree with me, but they will resent this intrusion. I have destroyed a brief moment of belief, and they are too few, these moments, and they must be savored. It does not matter that the floor under these moments of belief is idiotic. They need them; we need them. We need to believe in order and structure. And this is fine. Except it can get you killed in this place. Truly, this region, whatever we should call it, is now the cutting edge, and the old ways of seeing are being violated and scrapped before our eyes. We live in a fool's paradise, and it is fun to be a fool—come with me to Tuna

Country and I will show you fun—but it is also fatal. We do not live in the provinces or in the outback or in the wastelands. We live in the future. We are the laboratory where the future is being made. We must give this notion a moment's thought. Consider this: The think tanks are clueless; the presidents are living in caves and cannot see the sun. The experts are eunuchs. The real research is going on in Tuna Country, night after night, body after body. And we can do nothing about it. It is too late. The experiment has grown too huge to be stopped, the petri dish is shaking with energy and growth and wondrous mutant forms. The *chupacabras* do not exist. Tuna Country does exist. We must go to Tuna Country before it comes to us.

I am eating cazuela *and dipping a corn tortilla into the broth as he explains things to me. These places, he says, that I took you last night, well, how many times do you think I have been in those places in the past ten years? I'll tell you, three, maybe four times. I do not like these places. They make me sad. I took you there because I think if you see them, then these images will pile up on you, and you will make something out of what I show you.*

I nod and continue slurping my soup, sipping coffee, and tearing off pieces of tortilla. In the parking lot, an entire family is exiting the restaurant to their car, all wearing blue T-shirts that say "MARCH FOR NUMBER ONE" and feature a head of Christ. I pay attention to T-shirts since that is the way people now say what little it is that they have to say. And I agree with his argument about the images and the need to stuff my head with them.

I may even overdo this part of chores. I have these photographs that I seem to have failed to mention. A whore dead on the bed, blood glistening on her naked skin and the face at peace from an overdose. A hand reaching out of the sand of the dunes, the man executed and then buried with his paw waving at the uncaring sky— a signature of some devotees in the local mafia. Girls raped and killed, girls just killed, boys killed, lots of killing. A face blackened by the sun, lips curled, teeth white and prominent—sixteen or eighteen years old, I have forgotten—raped, murdered, dumped. Dozens and dozens of these images. Hardly anyone wants to see them.

I show the hand in the sand and am told it will cause nightmares. I mention some of the other images and am told to stop, such things will unsettle the soul. I go to a university press, squander two days, and then learn there is no market for such material. And besides, I am warned, it is unfocused. I confess: I do not yet have a photograph of the chupacabra—*night work is difficult even with strobes and flashes. But it will come. The images get sharper, harder, bloodier, warmer, stickier, and besides, time is on my side. My working schedule is this: the inevitable.*

God, it is hot as I squat by the thin shade of a creosote bush in Anapra, maybe thirty feet from the line and fifty yards from two border patrol Blazers waiting to snatch up some wetbacks. Around me in the sand are the prey. Two women trying to cross into El Paso so they can beg for coins in the saloons. There's a couple with two small children who have made it more than a thousand miles north from Oaxaca. They are short, dark, and Indian and have no place to go except forward. Off to the side is a group of men from Durango, who have just beaten off a gang of local thieves who tried to rob them. The talk is simple . . . it has not rained in my village in three years, there is no money in my town. . . .

The American officers in the Blazers are part of a program once called Operation Hold the Line. The government keeps hiring more and more of them. It is all governments seem to know how to do. A few yards away in one of the cardboard shacks of Anapra, I asked a ten-year-old boy what he wanted to be. He said, "A doctor."

I'm standing in the largest border community on the surface of this planet, a web of flesh called El Paso/Juárez. I am standing by the most crossed line on the surface of the earth, the U.S. and Mexican border with 232 million souls skittering over it last year with papers, and no one is quite sure how many made the journey without papers. It is supposed to be the only place on the planet where a First World nation rubs up against a Third World nation. It is also supposed to be a place that actually exists.

We'll see about that. And very soon I think because I can feel in my bones that rains are not coming. I can taste the ache in the blue of the sky. I believe. I've got a friend named Luis Urrea who explains it this way: The United States is a house and the border is the back porch of the house. And my friend thinks the back porch is on fire. I have no trouble with the claim about the fire. All my doubts center on the rain.

It is late now. The club La Serata is very fine with its marble facade and huge brass doors. The captain and two assistants guard this portal against the unseemly creatures of the Juárez night. The establishment has only been open a matter of months. The mayor attended the gala of the launch. Through some oversight, I was not invited. At the gala, the rich of the city gathered and touched the fine woods and marbles and gazed with pleasure on the reproduction of one of Michelangelo's frescos, the one where God reaches out to touch the flesh of Adam and electrocute him with the gift of life. At this opening, the proprietor sat at his own table drinking good liquor and snorting lines of coke. Life is not easy, I suppose, for Amado Carrillo Fuentes, the boss of the Juárez Cartel, and he needs his club for solace and decent companionship. He is credited with ordering the deaths of four hundred people in this city, but such numbers cannot be trusted in the flurry of events, and Carrillo is much too busy to keep decent records of such matters. But now as I stand at the door, I must deal with my own problems. I can hear the hearty voices of patrons from within, but the captain does not approve of me. To begin with, he notes, I am wearing sandals. Then there is the matter of my Levi's. And finally, he softly adds, my shirt has pockets. These are all transgressions against the club's dress code. He is sorry, but La Serata is a special place, an island of peace and taste, and surely I would not wish to violate its sound mores. So I am rejected, cast out into the midnight hour of a Juárez Friday night.

There is nothing to be done but go to the Zona Rosa, the district of whores, drugs, drunks, gangs, murders, rapes, and joy that shelters against the cathedral of Juárez. Many of the saloons here have a sign on the door that says, "No Cholos," but this seems to be the only form of discrimination. It is here that the police say a club of sociopaths gathered at a saloon called Joe's Place and snatched young girls and took them to the dunes by Anapra for their various pleasures. The leader of this band of revelers—once again according to the police—is an Arab. They are all now ensconced in the local jail and diligently tortured. Their confessions are a given. Once, they flourished here and would rape the girls, sodomize them, torture them, and then strangle them. As I walk the night streets, none of these recent frolics has left a trace. Tuna Country, a disco where the merry men once gathered, is in full roar with a wet T-shirt contest. The same is true of another of their haunts, the Fiesta. Joe's Place is temporarily dark, the doors pasted with police notices, but this is the only sign of their gregarious passing. I bump into a drunken Mexican whose T-shirt announces, "RED, WHITE AND BLUES." The atmosphere is much more welcoming to a wayfarer than the snooty codes of La Serata. A whore with bleached blond hair, a purple miniskirt, and a big stomach that reflects a zesty diet smiles at me and melts the night.

The air refreshes with its scents of roasting goats, urine, garbage, and the perfume of a passing woman. I walk past the cathedral, but it is very quiet and God seems asleep. After all these thousands of years with consultations with us on desert mountains and various messages on stone tablets, he may have had his fill. Perhaps, he has grown tired of us and left us to our own devices. I think this is quite possible. I don't believe there is anything else for him to teach us or show us. What is it that we don't know or understand? We must get down to business and make a good job of it. All the elements are at hand. And it is growing very warm, and I am more and more certain the rains will not come to us. I, for one, have no more questions. Or complaints. Considering the entertainment, the price of beer is quite reasonable. We've pretty much seen it all, except for the next pattern. I scan the skies but fail to locate the *chupacabras.* Perhaps later they will confirm my faith like the deer, the javelina, the coyotes, all my fellow creatures great and small.

Listen, hear the flap of wings in the night? They are sucking blood right now in the *maquiladoras,* plunging needles into eager arms in the byways. Money is streaking across the heavens, tirelessly roaming through wire transfers, and seeking a home at more than 20 percent. Amado is snorting at his table under the eye of the Renaissance God he bought and paid for. The cattle fall dead with a thud in the dust, their swollen tongues lolling at us obscenely.

We are drinking as the night thickens and clots. The Agent is very helpful and knows everything that is evil and illegal and kept secret from us. He is very broad shoul-

dered and has the soft voice and easy smile of a large man. The gun sits on the bar-room table between us and sleeps in a leather pouch. The man is careful to sit with his back to the wall, he has obviously read, as we all have, about Wild Bill and taken note. Of course, Mr. Hickok got whacked anyway. The Agent is happy as he tells me of the night they capped five or six of the fuckers, put guns to their skulls and rearranged their brains. He is a warrior and can face the war at times. At other times, he cannot, and tells me that after one bloody binge of two and half years he had to get out and smell the roses for a while. I ask about the legend, that the commandante *interrogated folks with a bolt cutter. He looks at me with mild scorn, waves his hand, and dismisses the question by spitting out, "They all do that."*

I am conditioned to being the innocent and the fool, and I take no offense. I am here to learn, and what I want to know is about the woman in the tambo, *that body floating in two hundred liters of acid as it coursed down a sewage canal in a barrel. She was taken from her fine home at 4:30 A.M. on the twentieth day of July—the anniversary of Pancho Villa's murder—by fifteen federal police officers or fifteen armed men that looked just like federal police officers. She had run a high-class club called Top Capos, the top bosses, and paid twenty to thirty thousand a night for entertain-*

ment flown up from the capital. She had an eight-week-old baby at the breast, and she was a widow. Someone had come for the father on May 3, and dad showed up again on May 4 beaten, dead, and bleeding from the rectum. I have tried to learn what I can. I have talked to pathologists who have put in those professional calls to colleagues—it turns out the FBI has someone who studies up on nothing but rates of decomposition. This has not helped—I must know the kind of acid and the purity of the acid to establish how long she did the breaststroke in that barrel. Besides, there is very little knowledge of such disposal techniques in my country; we have too much land to bother with acid and simply bury our kills or, more recently, favor wood chippers. The British, the experts tell me, they are into acid, what with their crowded lanes and festering mill towns. Everyone has tried to be helpful. It seems the acid was probably sulfuric, since the flesh was eaten but bones and hair persist-ed. All the experts seem to take note of the durability of breast implants, and I can see this odd fact popping up forever in their reports and scholarly papers.

And I have traced, as best I can, the genesis of this barrel floating in the sewage ca-nal. It seems one of the Agent's colleagues shared a list of informants with the federal police on the other side, and that's when the kidnappings and bad killings started to flare up. The snitches got snitched off. I mention this to the Agent, and his gaze does not falter, his expression does not change, and his voice stays soft and smooth as he says, Umm, could be. You see it all happened over there, on the other side, across a boundary, on another plane, in a place that cannot touch us here in this bar and can-not interfere with the pleasure of the cold beers we clutch in our hands. The world can be like this and often is. We can insist on order, we can have definitions and cate-gories and a sense of progress and security. Just stay out of Tuna Country. We can lis-ten to soothing music, wonder about sporting events, eye the women slinking around in this saloon. We can do all this. Except keep the gore from splattering into our faces.

Sunday comes down hard. I am sitting at a table in an American café, and my body is wracked with pain. A week ago I was in the boiler room of the desert and forgot to drink the appropriate amount of water. I was, as is my custom, savaged by a kidney stone attack. The boulder passed, and I went on with my toils. But an infection has set in—this sometimes happens—and now I make notes with my body electric from the thud of pain. But my penmanship is steady, so not to worry. Off in Anapra by the line, the village of cardboard houses grows at thirty to fifty shacks a week. Amado Carrillo is said to be ab-sent from Juárez at the moment, but do not worry, the boss is still in charge of the heavens. The *chupacabras* flutter around my head like prayers, and I have sent them off to Casa Rinconada to greet the new age worshipers who gather in the big kiva.

There is no rain or hope of rain. I predict the drought will last one hundred years, and when it ends we will all be a new people and we will dance and celebrate. I am saving up, putting pennies and centavos in a jar, for the fiesta when the drought ends, and trust me, I will be here if it takes forever. I am a creature of the long haul. *Chupacabras,* ah, they are transient, they will never last. I will last because I will never leave, not alive, not dead. So there. The daily paper prints a map of the barrios where water will be cut this summer in Juárez because the skies remain barren. A photograph on the next page shows mounted Mexican police in the desert looking for bones, the bones of young girls. Tuna Country.

A friend in Juárez set me straight on this matter. He said, "We don't live. We just survive." We are all borderline cases whether in Tuna Country or in the palaces of the governors. I'm a borderline case, but that's okay, I know why I am fucked up. To be honest, I kinda wonder about you.

Photographer Doug DuBois wishes to thank Rafael Cota for the use of some video images in the strips above.

Doug DuBois

Ruby Crystal Squid Dancing on the Rings of Saturn:
An Interview with Jaron Lanier

David Rothenberg

Jaron Lanier invented the term *virtual reality* to describe the place where people would soon be working and communicating with one another through computer technology. He cofounded VPL, the company that invented data gloves and helmets that extend human perception directly through machines. After leaving corporate life, Lanier is now a lecturer, consultant, and performing musician who specializes in the world's most exotic instruments, from the *shakuhachi* to the *khaen* to the *gu zchung*. Musical instruments, he has said, are often the most advanced technologies in the world's many cultures because they allow us to express such emotion and art directly that would be impossible without the tool.

Lanier doesn't really *like* computers. Why should he? (Science fiction writer Bruce Sterling recently said that no computer in use today has a life span greater than that of a hamster.) Lanier has a hopeful picture of the future of technology, but it is no blind techno-optimism. He often appears in the media to insist that new digital technology must not lead toward a dehumanized world, but instead foster new and more creative ways of communication between people. Technology, according to Lanier, will not save us, but it may allow us to shape collaborative dreams and inventions with a directness beyond language and an immediacy beyond symbols. Knowing how much he believes in a positive future for this planet and our species, I decided to speak with him about what the virtual world has to do with the natural world. Is there any place to touch the Earth as we wire up the senses for cyberspace?

Nature and Artifice

Terra Nova: Just what is virtual reality? And why should anyone interested in nature be concerned with it?

Lanier: Virtual reality starts with the idea of creating an alternate place by using instrumented clothing, such as goggles and gloves, to provide the human sensorimotor system with the stimulus and feedback that it would receive were it in a completely different environment. Virtual reality therefore becomes the first new objective place since the physical world. Multiple people can be together in an alternate environment, so that they see each other. Since they can see each other it means they must have bodies, and since they are in virtual reality it means they have to invent their bodies. A new kind of craftsmanship is required to create these bodies and the things that surround them. Virtual reality in general is at its weakest when you attempt to simulate experiences in the real world and is at its strongest when you attempt to use it as a tool to express unbounded imagination.

Terra Nova: What do virtual environments have to do with the natural environment?

Lanier: An environment is a thing outside of you. Reality implies much more. What is interesting about virtual reality is not so much the simulation of environments, but rather the exploration of aspects of communication that are possible if you can change the nature of the human sensorimotor loop. I am interested in modifying the human body so that a person could become a gazelle or a crab, from a sensorimotor point of view, controlling the body of a different shape with different responses. I am also very interested in people communicating within an objective world that is more controllable than the physical world.

Terra Nova: There seem to be two very different kinds of simulations you are interested in. One is the simulation of the life experience of real living creatures, and the other is the creation of artificial, different, but still human ways of interacting. What is the essential difference between these two kinds of invention?

Lanier: We have to recognize that we are *epistemologically impoverished.* The way we are in the universe does not give us many advantages in understanding what is going on, and we have to simply accept that. When I talk about nature and natural things, I am thinking about that *fundamentally mysterious sea* that surrounds, that we attempt to know through empirical methods but can never truly know.

When I am talking about something being artificial, I am talking about something that only exists through our personal and cultural understanding of it. Artificial things are things that exist solely in a cultural context. Things that are natural are those things that refuse to go away in any context—those things that are so stubborn that they must always be there and yet can never be fully known. Ironically, it is possible to say things with a level of precision about artifice that can never be said about nature.

Terra Nova: So an important part of something being natural is that it is more than human. There must be something that we cannot understand about it, that is beyond us. And if we claim to understand it all, we have taken away its naturalness because our methods of knowing are impoverished. Do you think that there are other beings that are less impoverished than we are?

Lanier: (*Laughing*)   That is a very interesting question. I think that all of the life- forms we know about have exactly the same problems as we do should they choose to look into these things. Though one of the fantasies I have had lately is that if I were a higher cephalopod, I would be less confused.

Terra Nova: If you were a giant squid?

Lanier: Right. Because some of the cephalopods have a few advantages over us. I have been very excited lately about their ability to create colors and patterns on their skin at will in a way that appears very emotional and very meaningful. It appears that cuttlefish, octopi, and squid communicate to each other with a color and pattern language.

Terra Nova: How do they make these colors?

Lanier: Part of the central nervous system of some of the cephalopods appears to be directly connected to colored cells (called chromatophores) at the surface of their skin that can change appearance very rapidly. The cells act like pixels in a great display. And they have enough variety of colored cells to choose from such that the range of colors and the range of moods, and of motions of patterns on them, is just fantastic, especially in the cuttlefish. There seems to be a direct link between their central nervous system and the color instrument on their skin, a very broad connection that is practically one of identity. So their mental state simply shows on their surface. Of course, at this point this is all speculation.

Terra Nova: Do they really look at each other? How do they respond to the colors?

Lanier: They look at each other, and other species are looking at them. Cephalopods all have extraordinarily advanced vision systems. They are concerned, among other things, with camouflage and the confusion of prey. But our problem is people, and I think that one of the reasons people become easily confused is that we rely too much on symbolic communication. And the reason we have to rely on symbolic communication is because there is an asymmetry between what we can perceive and what we can create easily. So we are very good visual perceivers, but in order to create a wide variety of visual information we have to craft artifacts. That takes much longer than the cephalopods' willing of images into being.

Postsymbolic Communication

Terra Nova: You have described communication through virtual reality as a movement towards *postsymbolic communication.*

Lanier: I usually describe postsymbolic communication by starting with a description of early childhood, which might or might not be correct. Young children experience an enormous fluidity of mind and an enormous innate creativity and inventiveness. But there comes a very shocking and disheartening moment in a child's life when they discover that rather than being the king of the universe—a fluid universe in which imagination rules and everything is different every day—they are instead helpless little pink, pudgy things in a completely different layer of reality. This other layer of reality is the physical world, and it is to this world they must turn to find their parents, their sustenance, their food, and companionship of any kind. But they are weak in this world. The products of their imagination cannot be realized in anything close to the time frame in which they think of them. So they have to gradually struggle between these two worlds, and eventually they either choose to live in the objective, physical world or else they end up in an asylum or in a rock band or something worse.

In our adult lives we still are trapped between these two worlds. In our dreams we can enter this subjective world of imagination in which anything can be true. But in our practical lives we have to leave it behind and work in this very compromised world where buildings are about the same every day and we are consistently weak, we can never fly or turn into cuttlefish or anything too unusual.

Now this presents an immediate problem: How do you communicate when you are so weak? We have devised a trick as a species—symbols. A symbol is an arbitrary way of using the very small part of the universe that you can control at approximately the rate at which you think to refer to all the contingencies of the universe that you are not powerful enough to bring into being. Little kids learn early on that they can control their tongues and their mouths and their hands and eventually the rest of their body, so the body is the part of the universe that you can control as fast as you think. And so we have this arbitrary set of contortions of these parts we can control to refer to all of the other things.

When I give lectures about this I typically make up an odd sentence, like, "All of you in the audience have just turned into ruby crystal squid dancing the samba on the rings of Saturn." Then I will point out that I have just saved trillions of dollars because to actually turn them into ruby squid I would have to spend a lot of money on genetic engineering and accelerating the space program and creating new life-support systems and even samba lessons. By simply flicking my tongue I have saved all that effort, so there is the magic of symbols.

Postsymbolic communication is the notion that someday, probably something like a century from now, a generation of kids might grow up with access to good virtual reality equipment and, even more importantly, to very high quality software tools. If they grow up with them they might develop enough fluency with those tools that they could start creating and even improvising the content of their virtual worlds, including both the objects in the virtual world and everything the objects do. If they were all doing that together, they might evolve a new type of conversation in which they directly created the content of an objective world instead of referring to it through symbols, cutting out the middle man. That is what postsymbolic communication would be.

Terra Nova: How about this picture of kids growing up, speaking of squid and the sea and fish and things: Imagine one kid who lives in a little village on the coast of Maine and grows up with the water, plays in the water and is very comfortable with water, goes swimming, goes diving, goes sailing. Learns about the water. Learns how to catch fish. Has some sense of connection with these creatures and knows a lot about the immediate environment of the sea through direct experience as a kid growing up.

And imagine another kid who grows up watching a lot of TV. He loves nature programs and he watches all these things about life under the sea. He learns the names of hundreds of fish living far deep in the sea, creatures that very few people actually get to see. Lots of information can be learned about them by watching TV! He watches giant squid at close range. He may go on to read detailed things about mysterious glowing creatures thousands of feet beneath the surface, or swim through the internet alighting upon information. He learns a lot through symbols about this fascinating world. He becomes an expert by age twelve. But you know what—he lives in Nebraska. He's never even seen the ocean.

Which of these children knows the sea better? And how would your new technology affect either of them in their future way of dealing with the world?

Lanier: Experience is primary, and the moment we lose it, we lose a great, great deal. My old slogan, that I continue to use as a guide in these matters, is "Information is alienated experience." If you know the world only through human artifice, you do not know anything at all. There is an insidious quality to that because you believe you know more than you do, and you create a self-fulfilling prophecy in which you do not know what could have been. It is exactly that process that leads to the sort of encroaching blandness of a lot of the twentieth century, the denaturing of everything, the sort of continuous turning to these human artifacts that do not have any meaning standing by themselves.

It is critically important to know nature, and that is why it is so important not to destroy it. But at the same time there is absolutely nothing wrong with human artifice; it is just important to be clear about what it is.

One of the things I like about virtual reality is that when you come out of it there is this wonderful experience of coming back into the natural world. Because your sensory systems become adapted to the simpler world of virtual reality, when you come back into the natural world it feels more natural than ever. That is probably the most valuable moment in the whole experience.

Terra Nova: When it's over? Do we need thin, virtual experience to appreciate thick, real experience? Sounds like coming down from an acid trip.

Lanier: No, an acid trip is stubbornly there—alas, it is natural. You craft virtual reality, and if you "space out" and stop doing anything with it, the illusion disappears.

Who Needs Plato?

Lanier: Let's talk about the problem of categories. When we use natural language to communicate with each other, we need categories, such as "red," to organize information for each other. If we were communicating postsymbolically, we would have an alternative. I could peer into my own past and pull out a familiar virtual object that I would show you. I would pull out a little pot, but if you look into the pot it is infinitely large inside, and what it has in it is every object I have ever seen that I consider to be reddish. Now if you look at that very large collection of objects you can feel for yourself what they have in common. It is a fuzzy kind of inclusion. I could even have multiple pots that intersect each other gradually, so that I do not have to make a sharp definition of what is red or what is not. But you empirically, directly can feel what seems to be in common about those objects, so now I have an entirely new way to communicate that is better than a category because I am sharing with you the direct experience of similarities among things instead of creating an abstraction. This very concrete style of communication is postsymbolic communication. If you combine extreme flexibility with rigorous concreteness you can end up actually getting rid of a layer of human artifice and abstraction—the category.

Terra Nova: Are you saying that there are no categories anymore in this way of presenting information?

Lanier: Well this is a new thing, and I do not have a name for this yet. I am stuck for now with language. It is better than a category because it is not an abstraction. It is based on your direct perception of a concrete large set of things.

Terra Nova: You open the virtual red pot, it's infinitely deep, and it's full of red objects. Aren't they arranged in some sort of way? Haven't they been organized? And isn't that organization somehow categorical?

Lanier: Well, only a twentieth-century person would interpret it that way. But the act is fundamentally different because I have created a relation among these things without having to imply a platonic redness. So I have gotten rid of Plato.

Terra Nova: Lots of people claim to have rid themselves of Plato, but he still returns. You're still presenting all these red objects as belonging to a category. And isn't the infinite pot a symbol of some kind?

Lanier: No, it is not. It is a practical object. It needs no name and need not be interpreted as being like anything else. It is a tool.

Terra Nova: How is this virtual attempt of communication different from a way of communicating where you and I are sitting in this room and instead of talking to each other, we walk around picking up objects and showing how they connect to each other? Without talking about them. Without claiming they stand for anything else. I walk around and I touch the cups over there and do things with them. Is that postsymbolic communication?

Lanier: It is postsymbolic *baby* communication because we do not have enough power over the physical world to truly communicate much that way. If we were more powerful in the physical world, we could communicate that way all the time and we would never have started using language. But here we are not. In the virtual world we are.

Beautiful Communication

Lanier: Let me give you an example from music because I have always found much clarity in thinking through music. The reason that synthesizer music does not tend to be as meaningful as acoustic music is not because it is made of electrons instead of molecules of air when it is produced, but rather that it is made of ideas instead of nature. So when you are playing a clarinet or piano, the instrument itself does not have any model of what a note is. The note does not really exist. It is a purely interpretive idea, and what is really going on is that as you play an acoustic instrument, you are reperceiving nature again and again, testing yourself against it in exactly the same way that a good scientist using empirical method has to. Whereas when you work with computers, a computer program has to embody cultural ideas in order for us to perceive it, and so therefore the idea of the note all of a sudden becomes something real from a functional point of view. Notes never existed before computers, and as soon as notes really exist, then music is lessened.

Terra Nova: Notes did exist before computers.

Lanier: I would argue they did not.

Terra Nova: What about composers writing musical notes, talking about notes?

Lanier: They were purely interpretive.

Terra Nova: You would say, then, that the note was a symbol for something?

Lanier: Yes.

Terra Nova: So the note was a symbol, and now through computers, notes instead of being symbols are the thing?

Lanier: Right. Our building blocks. In the Western musical tradition we have notation, but we have no way of knowing how much of music is really represented by our notation. Epistemological impoverishment once again. Ultimately I have a very strong suspicion that our music is much less represented by notes than we generally assume. Contemporary earwitness accounts of the music of Bach describe a music that would appear to be much closer to Klezmer than what we think of as Bach today—a sort of *swooping emotional wild.*

Terra Nova: But on the other hand, this does not bode well for the idea that a computer-controlled environment brings you beyond symbolic communication. You seem to be going backwards instead of forward into symbollessness . . .

Lanier: No. Actually, I believe that if you go all the way using information systems, communication will be better.

Terra Nova: I see.

Lanier: Going halfway is the problem. I believe that natural language is a halfway step, confusing the matter. If you go all the way to postsymbolic communication, then you have a situation of improved clarity.

Terra Nova: You think that it is somehow possible to have deeper and more evolved and more efficient or more mysterious kinds of communication through computer technology?

Lanier: I think it is possible to have more beautiful communication. Can I address this question from a different angle which is more political?

Terra Nova: Sure.

Lanier: At some point Western culture embarked on a project to become more powerful than nature and to do that methodically, to embark on this ramp ever upward. The original rationale for this was that nature was scary and dangerous, and in fact it was. Nature meant mysterious diseases that were terribly cruel. Nature meant volcanoes, and so forth. By the time we hit the twentieth century, what has happened is that science and technology have gotten so good that our main problems come not from what we call nature, but from our own behavior. So the situation is fundamentally different. The question is why are we still making new technologies, and why are we still trying to find new science? There are a few cases where the original agenda is still valid, such as trying to understand the cause and cure of AIDS or trying to understand earthquakes, but almost all of the science and technology we do now cannot be justified on those original terms.

I think a technology can no longer be judged purely on the basis of whether it is clever or whether it makes people more powerful in some way or other. I think technologies now need to be judged also in terms of what metaphors they provide to society at large, what it feels like to use them, what influence they are likely to have on culture, what McLuhanesque side effects they will have. I think a technologist who does not work on all these levels at once is simply being irresponsible.

Terra Nova: So you are advocating more ethics, more responsibility?

Lanier: More artistry, really. I think that this should not be framed from the negative; it should be framed from the positive. I think it would be disastrous to have some sort of humanist police running around telling scientists what constitutes good research because that would create horrible, cowering science.

Terra Nova: Though you are asking the scientists to learn to do that themselves in some sense. It is not just making science more like art, but also advocating some kind of responsibility, don't you think? Art sounds more fun, but you are also saying that going for more power is not the goal.

Lanier: Real art is responsible. The problem with art today is that it has become a token for pecking-order exchanges between people in various forms— whether it is in the sort of prestige of collecting what is considered legitimate art, or whether it is in the status fantasies inherent in rock music, or whether it is in the form of socialist-realist artificially created and constrained art. I think that there is irony within irony in all this. This brings up a quite different side of nature which has to do with some of the ugliness of who we are—our old animal business showing through. The truth is that nature is cruel from a rational perspective, and so much of what human affairs consist of does seem to be these horrible bits of business about maintaining pecking order and using various forms of violence to compete.

Terra Nova: There is still a moralistic tone to what you say; it is not all aesthetic. People who worked in previous technologies probably did not think enough about what their work would lead to. The atomic bomb, television, and the automobile were all built with excitement. They sounded wonderful; they offered seemingly limitless possibilities. Did we have to make all of history's mistakes?

Death and the Internet

Lanier: Here I think that there is an element of Buddhist philosophy that should come into play: Ultimately, intention has to matter. It is really impossible to know what the effects of actions will be. It may be easy to blame a technology for a problem, but it is much, much harder to try to come up with a real balanced assessment of what the alternatives might have been.

If you try, if you attempt to do things that are good, I believe that ultimately things will be better, if not perfect, as a result. I find myself in very sharp disagreement with a lot of the extropian folks now who believe that trying to affect things is the worst thing, that we should instead let some sort of evolutionlike process consisting of capitalism and science and technology communities compete in an aggressive fashion. They believe that any intervention we make will make the situation worse. I just reject that way of thinking.

If we abdicate the role of human agency, even with all the ambiguities involved, then we are completely lost. Then we have really given ourselves up, we are just hoping to become part of some big computer. If I really believed that computers and information existed without us, I might be willing to take that risk, but since I think that they are only a fantasy, to me it is practically suicide.

When the automobile was invented, we were still in the zone of needing more power over nature. The truth was that we needed a way to get around without having to lay down railroad tracks and without having to walk or ride there on a horse or something. There was actually a feeling of inadequacy compared with nature. I do not think anybody has that feeling with regard to transportation today. I think we all feel powerful enough to be transported, with all the available technologies. Now the real question is aesthetic: How much misery, smell, congestion, and absurdity are we willing to accept in our transportation systems?

People want their cars so much for existential reasons: It is my car, I am going where I want, it has my shit in it, it expresses me. I think that is a kind of a nutty notion, fueled by advertising myths. But you can still understand that desire, that feeling—at least in certain cultures—of this mind frame of individual power and individual attention. Now what I am hoping is that as media goes interactive, we will start to see with the Internet that same existential need, fulfilled with media instead of actual physical excess.

Media have never before allowed more than a very small number of people to be themselves in this sense. The Internet, specifically the World Wide Web, is also the first working anarchy in history, and so I think it says something surprisingly positive about human social potential. When we are able to act without concern for dividing up a limited pot of resources we are capable of cooperating. That is very nice. The Internet also does a very peculiar thing—it creates a simulation of the mysteriousness we find in nature. I like to describe it as an erotic quality. If you are on the web it might be quite dull but you never know what may be around the corner, so you are drawn on to try to find what might be there. Perhaps the Internet reveals a most natural aspect in ourselves that previous media have not.

Terra Nova: Do you mean offering accessibility to the mystery? Or mirroring the mystery in its structure?

Lanier: Most media in the past have tended to create a proscenium arch in which we put on a show of how we think we should be. The first medium that was able to allow us to be as we are was the telephone. The problem was that it was a show put on only for God because each of us only perceived the telephone call we were on, so the great wildness of simultaneous telephone calls has remained completely mysterious. But with the Internet, we can travel among conversations. We may discover each other in a natural way simply because so much is revealed at once in such an unplanned fashion.

Terra Nova: You would say more so than writing letters?

Lanier: Yes, I would say that.

Terra Nova: Because it is quicker?

Lanier: No, because it is collective and publicly exposed.

Terra Nova: That is true. There is this mundane, practical aspect of an actual world where people may communicate through the machine at home. Then you can live where you want and work from home. The Internet, or virtual reality, does not tell you how human settlements need to be set up—it leaves environmental design open. It could stay the way it is now or people could have time to set up different kinds of communities. It does not prescribe anything there, it just tends to free us from physical constraints. That could help, but it might not.

There has to be a way where this new technology does not just divorce people from nature. You say that *artificial* and *natural* are two different categories. But people and the world have to be made more connected.

Lanier: Sacred things are visited infrequently. If you spend all of your time in church, it is no longer a church. I think that our only practical hope is to treat nature that way. Considering how large our population is, there is not enough to go around. I think that is very sad. On the other hand, I think it is realistically the only practical way to think at this time.

Terra Nova: That is only one very specific sense of nature—as a beautiful place that can get ruined when there are too many people there. Another view says that we have to live in such a way that we are brought in daily contact with the world that is not human so that we respect and really work with it so that we learn where our food comes from, how our waste is used, and how our energy is used up—so we do not hide from it and live only inside technology. I think that you may be right: We should not overrun the same places. Do you really want more separation between humanity and nature?

Lanier: Even the most enlightened sectors of Western civilizations seem to spread like a blight into the world, and every little bit of beauty is destroyed by the very act of appreciating it. Every traditional culture, every natural place is lessened every year by the act of so many people being interested in it. I think somehow we have to visit those places that are different from our usual places less frequently, and we have to find more sustenance and more connection with fundamental reality where we are.

Terra Nova: Won't we want to visit them less frequently if our own place where we live day to day approaches the qualities that we tend to seek somewhere else? When we live in a place and we wake up and say, "This is a beautiful place to live and a fine way to live and there is nowhere else I want to go?"

Lanier: Busy minds need places to play. Certain things that are now done in the natural world should be done in virtual reality instead.

Terra Nova: Such as?

Lanier: Dune-buggy riding. (*Laughter*) I would like to make a buggy ride in virtual reality so wonderful that nobody will ever tear up a desert again. This is an example of a human expression of autonomy and power that might very well be equaled in the world of artifice, and I think there is a moral imperative to attempt to do it in that world, not the fragile desert world.

Terra Nova: How about hiking? Should that be done in virtual reality?

Lanier: Well, you know, these are hard issues. Speaking hypothetically it is easy to say, "Yes, we should all spend more time in nature." I am trying to speak practically and just recognize that if we look at the population we are likely to have very soon, these are going to become practical questions. It is not pretty. I am not saying this is what I want; I am just saying that we are going to be facing a crisis of overvisiting our own beautiful places.

Terra Nova: Would you say that going to a health club in Manhattan is closer to virtual reality than running out in Central Park? There are machines that simulate hills and changes, and you are meant to feel you are going through an environment. More people can do this kind of exercise in tight quarters, and when totally engrossed they can feel like they're alone.

Lanier: Well, you have to remember the point of virtual reality is not simulation, but enhanced expression.

Terra Nova: Yes, I understand that, but dune-buggy riding is probably not about enhanced expression as it presently is enjoyed. The people who want to do it—They want a thrill; they want to get outside and have a good time.

Lanier: It is a little like adolescence, veering as close to suicide as possible. Destroying nature is the most dramatic way of denying death and denying the scarcity of life, pretending to be infinite.

I think virtual reality makes it easier for people to perceive the existence of their own consciousness, their subjectivity. Because the objective world is so stark, you perceive yourself with increased clarity. I think that is one of the exciting things about the experience. From an absolute point of view, of course, I think that nature is more important, and I hope that virtual reality is really a tool for perceiving nature by means of comparison. I believe, in general, media technologies have served that function. I think that we can better appreciate the quality of natural sound by having artificial sound to compare it to.

Terra Nova: The toughest thing to bring across here is the sense that using this new technology really allows natural mystery because many people are going to think that a world where communication is mediated by computers is a world that is more artificial than ever before.

Much of the quality of today's technology must change. Machines must really bring people together and recognize differences, reaching out to people who cannot afford the latest toys. Solve the combined problems of lack of community, anonymity, alienation. You are right that information in itself is nothing. The universe is not made of information. It is not like Philip K. Dick's words set to music in Tod Machover's opera *Valis,* "The universe is made of inforMAtion!" Technology must prophesy something else than this.

Lanier: If people believe the universe is made of information, they think they do not have to die. If the universe were really made of information, then hypothetically you could hook your brain up onto a computer medium and you could live forever. But that is not reality.

It is this newfangled form of death denial that gives computer culture its frequently bland and nerdy quality. It is seductive to believe that everything is information. Not only might you not have to die, but you even have a chance of understanding and containing all of nature, rather than being subject to a big mystery.

But computers don't exist by themselves out in nature. They're culturally relative objects. They exist by virtue of our experience of them, and experience itself only exists because of our epistemological impoverishment. If we could know everything, we wouldn't be capable of surprise, so we wouldn't have experience, and then computers would cease to exist.

Terra Nova: I am getting the sense that you think nature is somehow the world *as it is.*

Lanier: Right.

Terra Nova: So I would just like to know: When have you personally most felt the world as it is? One or two examples.

Lanier: (*Pause*) My mother's death.

And there have been times when I have been in nature hiking, especially in Mexico. I think you perceive reality best when you perceive what you want to see and what you do not want to see simultaneously, in contradiction. Like seeing the incredible beauty of a bird of prey as it swoops down to devour.

Terra Nova: Death is an important part of the way things are.

Lanier: Perceiving nature without death is not perceiving it at all. Then you are in Disney World.

Wild Turkey

Ray Isle

It was 1987, and I was an outsider. After all, didn't I have the requisites? Check it out. I'd moved that year to Austin, Texas, a fine place for would-be outsiders. I'd found an appropriately shitty job, working as a "reservationist" on the 800-number for Sheraton Hotels. And I had the other crucial outsider qualifications: the Membership in a Bad Band. The Miserable Hole to Live In. And, crucially, the Substantially-More-Put-Together Girlfriend. Patrizia. Student of law. Beautiful, smart, German. And understanding—so far.

Despite all this, I felt overcome by the squalor and pointlessness of my life. Sure, I was an outsider. My fellow reservationists, mostly Air Force wives from Bergstrom Air Force Base, could vouch for that: "You're in a band called The Stumps? That's gross." "Who'd want to listen to a band called The Stumps?" Well, no one, except maybe Waxface Jeff, our roadie; but that might have been an act on his part. We were also his biggest customers for the lousy pot he sold out of the house we all lived in.

I was beginning to catch on. Being an outsider meant being no one. And given that Patrizia was about to graduate from law school and start making $80,000 a year, my no-one-hood boded ill. Our love was about to have oxygen injected into its veins by the assassin of financial incompatibility. Things, it seemed, were about to suck.

So, Friday afternoon in late July. I am sitting in my sound-baffled nook, inventing for a lawyer from Detroit the sublime glories of the Sheraton Bora-Bora:

"The beach? The beach is fucking gorgeous. White sand, acres of it. You climb down these curving wooden stairs from the hotel—which is on this cliff you've got to see to believe. The water? The water is beyond real. It's unearthly. Better than Yves Klein's 'Universal Blue.'"

"What?" The guy's a litigator. "Fucking" he understands; Yves Klein, no. "Listen, I don't give a shit about this Klein guy. All I know is I need a room. Two weeks, check out the fourteenth. Make it a suite."

Sipping black coffee, staring out darkened windows at the supraluminous natural world, I watch Patrizia cleave rapidly away through azure expense-account waters. I see clouds like Egyptian cotton pillowcases, sand like silk. And if I lean over slightly, I can even see myself. On the shore. In a little blue uniform. Raking up the monkey shit.

"Okay, this is what I want," the litigator says. "I want a minibar, I want a *king*-size bed, I want—"

I disconnect him. Hey, I'm an *outsider,* aren't I? Clearly it is time to get *out.* "Guadalupe Mountains?" Patrizia was puzzled. Whenever she was puzzled, she looked stern. "What are the Guadalupe Mountains? This is Texas. There are no mountains."

I explained. National Park. Just east of El Paso. Guadalupe Peak itself, highest point in Texas. Arid. Hostile. Rocky. Cactus. Mesquite. Gila monsters. Rattlesnakes. Bobcats—

"I have advanced torts," Patrizia said sternly.

"Gosh, Zipa, you ought to get something for that."

Patrizia laughed, so lightly that it wasn't entirely clear whether she was laughing at all. What was clear were the white, sharp tips of her perfect teeth. "Funny," she observed without a trace of amusement. She hated being called Zipa.

She said, "This is one of those male things, isn't it?"

"This is not a 'male thing.'"

"No, it's a male thing." She added with certainty, "You should go alone."

Well, *good.* After about five seconds of thinking about it, I realized Patrizia was right. This *was* a male thing. High time, too! No more of this namby-pamby camping-trip b.s. Camping trips were for families and tourists. I could already feel myself puffing up with maleness, like one of those colorful Amazon toads. No sir, if you are twenty-two, disaffected, and not in the company of your girlfriend, one thing you do *not* do is go on a camping trip. What you do is engage the wilderness one on one. You test yourself. You see what you're made of. It's a pre-Jesus activity, a been-to-the-edge-and-survived trip. It's starving yourself in a pit while eating hallucinogenic mushrooms. It's being Richard Harris in that movie "A Man Called Horse," where Comanches haul you up in the air by means of sticks stuck through your chest muscles, though I really wasn't keen on anything quite that extreme. In any case, you hunt down personal epiphany and wrest it from the bloody jaws of the unthinking wilderness. That's the general gist of the thing.

"You're right," I told Patrizia. "This is something I need to do. Alone."

"Of course I'm right."

"I'll miss making love under the stars, though."

"Making love under the stars is itchy. If you come back, we can make love in the bed."

Something was bothering me, though. "*If* I come back? What the hell's that supposed to mean?"

Patrizia shrugged. "You never know."

Five A.M., a sleepy good-bye kiss from Patrizia still on my lips. I am packing the back of my 1979 Ford Fairlane station wagon with the rudiments of survival: tent, lantern, sleeping bag, pillow, ground sheet, foam liner, another pillow, backpack, flashlight, rope, camp stove, propane canisters, Walkman, one hundred and thirty cassette tapes in two faux leather cases, paperback copies of *Moby Dick, Desert Solitaire,* and *Fear and Loathing in Las Vegas,* a quarter ounce of Waxface's best skunkweed, rolling papers, canned tamales, canned chili, canned beef stew, a canned plum pudding complete with hard sauce (my mother had given it to me for Christmas), cigarettes, jar of instant coffee, Styrofoam cooler, beer, more beer, a fifth of Famous Grouse blended scotch whiskey, bottled water, magnesium flares, topographic maps, spf 55 sunscreen, $150 in traveler's checks, running shoes, hiking boots, moleskin, insect repellent, oranges, a snakebite kit, several aluminum pots, a can opener, two ten-pound dumbbells, and a guitar.

Presently Patrizia joins me in the driveway, holding a cup of coffee.

"What?"

She shrugs.

"These are all necessities."

"I did not say anything."

"You were doing that saying-something-without-saying-anything thing."

She sips her coffee. "Have fun," she suggests. "If you kill anything, be sure to bring me back its head."

It was in the hundreds as I rambled up the access road to Dog Canyon campground. The Fairlane bottomed out every fifty feet. I'd decided to skip the south end of the park. Enough of this candy-ass forest shit. I wanted aridity, sterility, the rattlesnake slipping through the eye socket of the cow skull, the sun like God's disapproval. Baking salt flats. Million-year drought.

I had my shirt unbuttoned. I was wearing sunglasses, smoking a Camel. The wind was ruthless. Squinting against it, I felt weathered and tough, Clint Eastwood in high-tops and shorts. There was no one around. The hot, desolate wind cracked past empty picnic tables. I parked and got out.

The silence was immense.

Sheraton Hotels had never even *heard* of this place.

An hour later, I was sitting on the picnic table of the site I'd selected, reading *Moby Dick* and batting the persistent desert gnats away from my eyes, nose, and ears. Abruptly the silence, the immense silence, *my* silence, was rent by a rumbling. A wedge of large men wearing leather and denim roared into the campground on Harleys. *Jokers,* their jackets said. *Fort Worth Chapter.* Here I was, holding my book, wearing a baseball cap, sunscreen smeared on my nose like Crisco. If I'd had a jacket like theirs, it would have read "Lightweight" or maybe "Panty-boy" and been embroidered by my mother.

I decided to keep reading, to show my nonchalance. The Jokers disappeared to the far end of the campground. About fifteen minutes later one of them sauntered up. He looked like a thirty-year-old version of a kid named Randy Ray I'd gone to junior high with. Motorcycle boots, no shirt, jeans impacted with grease, long red hair in a ponytail, pockmarked skin, gray teeth. A stomach you could crack nuts on. Randy Ray had sat down next to me at lunch once, and rather out of the blue he informed me in his bolted-down rural Texas accent that "the best thing about them souvenir bats they give away at Astros games" was that you could "wrap yourself some bicycle chain around the big end, get some duct tape on there, make yourself a real good nigger-knocker that way." Then he'd whipped out a butterfly knife, removed my sandwich from my hands, whacked off half of it, said, "Thanks, dickweed," and walked off.

This adult version of Randy Ray flipped a glassine envelope onto the table next to me. "You want to buy some crank, man? Good shit."

Hey, I was a *Stump*, man. I knew how to be cool.

"Not today. Thanks anyway." I handed the envelope back to him.

He studied me for a moment. Then he palmed the speed and checked out my campsite. Nylon tent, backpack leaning against the car, Styrofoam cooler, guitar, dumbbells.

"What are you doing—camping?"

The amused, boot-to-the-head twist he'd given the word *camping* didn't make me optimistic. "Yes I am," I said.

He smiled. I recognized that smile: Bruce Lee smiles just the same way in *Enter the Dragon* right as he's crushing some poor fool's trachea with his foot.

Randy Ray said, "Come on over and party with us later, man. If you're up for it. We're gonna get fucked up."

E.L., the manager of the Guadalupe Salt Flats KOA Kampground, thirty minutes away on Highway 90, turned out to be a very sweet old guy. He even reduced my campground fee by five dollars after I explained that, no, I did not need an electrical hookup for my "recreational vehicle."

The following morning I drove to the forested end of the park. I was relieved, even as I felt irked with myself for feeling that relief. But here was normality. The gravel road wound through stands of piñon pine and gray oak. Pale clouds scudded through the hot sky. Birds sang. The air smelled rosiny and fresh. Ahead, the brown-painted cinder blocks of a ranger station rose, inviting, a battered drinking fountain by the door. The U.S. flag fluttered overhead.

"Hey, how you doing, buddy?" the ranger called when I went in. He was a little too friendly for my tastes.

I was doing fine, I muttered, and went through the business of registering for a campsite.

"Tell you what. I'll put you over there by the creek," he said. "That's real nice. One thing, you might want to keep an eye out for the turkey." He nodded, checking slowly over the form I'd filled out. "See, there's this flock of wild Mexican turkeys living around here. You might want to keep an eye out for the gander. Kinda thinks the campground's his kingdom." He chuckled, evidently amused by the turkey's territorial delusions.

"A turkey," I said.

"That's right."

With the same nonchalance I'd attempted yesterday, I observed, "Well, it can't be any worse than a motorcycle gang."

"Are they up at the canyon again? God*damn,* those guys are worse than fire ants."

Nothing is more irritating to false nonchalance than real nonchalance. Nevertheless, as I was leaving it occurred to me to ask, "So—what are you supposed to do if this turkey decides it's feeling aggressive?"

The ranger laughed uproariously. I was more irritated. We both waited for him to catch his breath. Finally he did.

"Oh, just make a lot of noise. Yell at him. Shoo him off! That old boy'll get scared, he'll figure out what's what."

Noon. I am pounding the stakes of my tent into the hard-packed West Texas dirt. Shirt off, sweating bare-backed in the sun. Testosterone pumping through every fiber of my body. This is it, maleness, solitude, *cojones grandes.* I feel strong. I feel powerful. Then I hear the gobbling.

It occurred to me, briefly, to pay attention to it. But really, who gave a shit if there were a hundred turkeys roosting in this campground? My run-in with the Jokers had left me analytical and cold and not very willing to back down. I had come out here to go *mano a mano* with the brutal truth of nature, after all, not to spend my time worrying about fat birds so mindless that if you leave them standing outside in the rain, they drown.

The gobbling ululated again, causing me to whack my last tent stake into a pretzel shape. "*Damn,*" I said and looked up. At the top of the dusty rise behind my camp, a large turkey had appeared. It was bronze in color, wings tipped in white. It held its ruddy, wattled neck and head high. It looked something like a rotund vulture, but lower to the ground.

I tossed the useless tent stake aside and stood up, wiping the sweat off my face.

The turkey paused at the crest. In a sort of Napoleonic moment, it took in the camp and saw me. Its head cocked to attention. Then it gave a particularly strident gobble and trotted down the slope.

My thoughts can be transliterated, roughly, as these: It's a turkey. Give me a fucking break.

I picked up an aluminum pot and my heavy-duty can opener. Make noise? Okay, you got it. I was going to scare the bejesus out of this bird.

As the turkey came down the hill I went up to meet it, banging on my pot and yelling, "Shoo! Shoo! Hyah!" That was when I learned my first lesson in Turkey Behavior 101. What I learned is that turkeys don't give a shit when you make noise. Maybe they don't have ears. I still don't know. I do know that as soon as I got near it, banging away and hollering, amused and very pleased with myself, the turkey leaped into the air. It battered me with its wings and raked at my face with enormous black claws.

"Jesus Christ!" I said, and ran.

As soon as I got about ten feet between us, I turned and started banging on the pot again, harder this time. "Hah! Shoo! Get out of here!" I yelled.

This time the turkey tried to tear my face off.

"Jesus *Christ!*" I yelled again, and ran.

We went through this three or four times. Then I punched a hole in the bottom of my pot with the heavy end of the can opener. "Uh oh," I said. The turkey quickly launched a brutal counterassault, buffeting me and gobbling and trying to crawl up my chest. I flung the pot aside and dashed to my car, thinking for some reason as I did, Good Lord, what if I get hepatitis from this thing?

Therapy can sort out why I felt turkeys might be a source of hepatitis. In the meantime, let's hold me suspended in air as the car door slams shut, and consider some of the facts of this situation:

This is late July on the Texas–New Mexico border. Now, the average temperature in late July along the Texas–New Mexico border is something like ninety-eight, ninety-nine degrees Fahrenheit, but this day in particular happens to be about one hundred and five. Even the flies have passed out. The car I have just vaulted into is a dark blue Ford Fairlane station wagon, circa 1979, with dark blue vinyl seats. It has been sitting in the direct sun, windows closed, for something like four hours. And I am wearing shorts and no shirt.

Soon after entering my recreational vehicle, I realized a moment of extreme lucidity. This was followed by pain. I made a horrible noise—something like a shrill gobble in fact—and then proceeded to levitate, a skill I had not known I possessed until that moment.

Outside, the turkey took up position on top of my picnic table and started eating my Doritos.

At that point I had my epiphany. I had come to the wilderness looking for an epiphany, of course, and though this was not the epiphany I had hoped for, it would have to do. What occurred to me was that even in the direst of circumstances, trapped in a car as hot as a pizza oven, nagged at by the thought that (a) you have just been outfought and even outstrategized by a bird most people consider holiday dinner and (b) back in the civilized world your friends are investing in mutual funds, your girlfriend is checking out Mercedes sports coupes, and you are still buying canned tamales for dinner—despite all this, the wonderful thing about nature is that poverty asks no comparisons there. Even that Detroit lawyer, sunning himself on his Sheraton beach in Bora Bora, could have his legs chewed off by a shark if he didn't watch it.

Whereupon a Range Rover drove into the campsite one down from mine. It parked, and a wiry, balding fellow of about forty got out and stood, hands on hips, observing the site he'd selected. Moments later a much younger, much taller woman with a lot more hair than he had (all blonde and rumpled up in a gosh-isn't-athletic-sex-wonderful kind of way) poured herself out of the passenger side of that forty-thousand-dollar vehicle. She flowed up to him, more liquid than mercury, and started nuzzling his ear. The turkey studied the two of them, then me, and in a turkey insight of startling clarity understood that epiphanies were bullshit. It returned its attention to my Doritos. The bald guy and his girlfriend opened the swing-back of the Rover and vanished inside. Within moments it began to bounce.

Sexual jealousy, I've found, often leads directly to inspiration. Crouched there, sweating, sucking in the superheated air, trying not to come in contact with any surface, I came to a conclusion. The conclusion was simple: This was *ridiculous*. This was *sad,* man. This was *pitiful*. This was citizenship in the country of the wimps. Randy Ray would have laughed in my face, then eaten my canned plum pudding. Hell, Patrizia probably would have, too. Get real, I told myself. You're a human being. A tool-using creature. And that thing out there?

The turkey, serene in its inarguable turkeyness, not hungry for existential justification (or even anything at all, since it had now finished my Doritos), settled down for what looked like an extended roost on my picnic table. Fuck you, I thought. If I'd had a shotgun at that moment, dinner that night would have been extravagant.

I had no gun. But—call it a second epiphany—it did occur to me that I had a brain.

I moved. The turkey cocked its head and gave a low warning gobble, but I was fast, scuttling from the station wagon, apelike and low to the ground, scooping up stones from the dirt. And as the turkey launched itself from table to earth, I hit it in the ass with a rock.

Success! It seemed worried. I hurled another rock, advancing. That's right, *bird,* know what you're dealing with? Monkey-boy ascendant! Opposable fucking *thumbs!* I whipped stones at it. It ran. There's one for terraced farming, pal, and one for irrigation. How about another, for Copernicus? What about the internal combustion engine? The turkey scooted up the hill, gobbling worriedly. I grabbed more stones. I was filled with righteousness. Screw nature, man, give me civilization! Plug in the amps, kick out the jams, let's hear it for calculus and philosophical inquiry, steel-toed workboots and heavy industry, freeze-dried hiking rations and Gore-Tex, distortion and the Fender Stratocaster.

The turkey? Gobbling in outrage, it topped the rise and vanished.

"You threw rocks at a bird?"

I was on the pay phone behind the ranger station, talking long-distance to Patrizia in Austin. "You don't understand, this thing was huge!"

"You had a fight with a turkey?"

"Patrizia, listen, the thing was nuts! It was like this giant, rabid bird!"

"You were frightened by a turkey and so you threw rocks at it? This makes you proud?"

"Well, yes and no," I said. "But it was really very big. Huge, Zipa. Prehistoric."

"Good thing you were alone. I might have shrieked and needed protection."

Ah, sarcasm. How I'd missed it. Patrizia would have needed no protection. She would have snapped the turkey's neck, roasted it over an open fire, then fixed us turkey sandwiches.

Dusk hung over us both and over the six hundred miles between us.

On my way back to the camp I nearly ran smack into the blonde girlfriend of the Range Rover pilot as she delicately stepped from the women's porto-can.

"You're the guy just one up from us, aren't you?" she said in her melted-butter voice.

I admitted this was true. Then I asked her if she was enjoying her camping trip.

"My *God,* last night we partied with this motorcycle gang, the Jokers. I was freaked out of my wits. But Alan, my boyfriend, he was *great* with them. They *loved* him. It was amazing. He's in criminal defense in Dallas. He deals with guys like them all the time."

No, no pride. But even if you can't squeeze pride out of the aged lemon of life, sometimes you can still recover a few small, sour drops of victory:

I am heating my tamales for dinner when out of the dark heart of the night the primordial gobbling rises again. It bubbles up like oil, undaunted and instinctual. And down on the flatlands the balding guy waves up to me. He's looking a little desperate. His girlfriend is huddled behind their picnic table. He's dodging and feinting as the turkey advances.

He yells, "Excuse me, buddy, but do you know what we're supposed to do about this thing?"

Panty-boy. Lightweight. "Oh sure, just make a lot of noise! It'll get scared and run away."

I figure five minutes. Maybe ten. More than enough. Then, hormones humming like magic, I can head down into that violent darkness and save them.

Shepherd, Transylvania, 1994

Romania

Rick Bass

Photographs by Raymond Meeks

You might find yourself in spring in the small, sleepy village of Varsag, where everyone is friendly to you. You could allow yourself to be as foolish as a grouse and think, "Ah, *civilized* people live here—what great peace, harmony, and friendliness!" You might then be surprised to read that it was only four years ago that a hundred thousand Romanian secret police roamed the country, the *Securitatae* of the dictator Nicolae Ceausescu. The specificity of the torture and terror: cutting open pregnant women. I cannot bring myself to elaborate on the atrocities. I have come here to write about Romania's bear population—to travel through the dark chaos and madness of history and humanity into the light-filled wilderness, and the peace that accompanies all things living with grace. If I were to tell you of some of the horrors that the Romanian secret police inflicted upon the populace, you would crawl wretched from this century; you would want to run into the woods and hide. I will not abuse you with horrible knowledge—or rather, with its specificity. But you must promise, in return, not to hide behind the dangerous cloak—racist, at best—that "they" are different. You could then travel to this country, where a smiling man might give you directions to, say, the Liberty Bell or the Statue of Liberty. And you might think, "This country *is* different, these humans are *better*." You might think that until you read the history of the Civil War between the States or of our slave trade or of our genocide of the native populations . . . I will not list tortures to the human body and spirit in my country, but you must promise to reexamine, reconsider, the safe premise that "they" are wild savages, that this could never have happened in our country, and never has.

I believe strongly that how you treat the land, or any animal on the land, is ultimately how you will treat another person. I also believe that all of these problems have at their source a lack of respect; that a bad person strikes out by abusing the land, which he or she believes, mistakenly, cannot strike back. Getting away with that, the bad person then mistreats animals, which have no voice, and gets away with it—exploits them for gain. And then he or she exploits minorities, who have no voice or whose voice the tyrant pretends not to hear. And so on. In many ways, then, the green rage of environmentalism is a distillation of rage to its most elemental focus: the inability, or lost ability, to treat *anything* with respect. How we treat the earth in our culture becomes the model for every relationship we have as we move across the earth to which we are inescapably connected. I believe, then, that my work as an environmentalist is good and right and has a feel to it of the sacred.

But then I land in a foreign country where the citizens are being cut into pieces and fed to the pigs; where blind, maimed beggars crawl the filthy streets moaning; where, in the cities, clouds of carbon black rest on children's lungs like ever-present fog; where the water and the grass and the sky shimmer with the poison-buzzing echo of Chernobyl; where centuries-old family homes have been bulldozed—destroy the individual! destroy the wild seed!—and replaced with spirit-gray Communist bloc buildings, all the same, same, same. I come flying my white, pampered ass into a place where the physical act of existing is a battle, and my faith and belief in what I am doing, in the significance of it, is shaken. Rocked, actually.

I question whether or not I have tinges of that evil, to come into a foreign country all hot and bothered about their *bear* population as I step over many of the citizenry who have nothing, who for a long time were reduced to being nothing more than a species of livestock.

It is springtime. The cherry blossoms burst white against the green hills, fresh from the light spring rains. Not yet leached by the poisons of mass agriculture, the fresh-turned fields are as dark as a black stallion. There may be centuries of bones beneath those fields, the soil may be rich with human blood, but it is spring again, communism is gone, and everyone has been set free; the humans in this particular landscape are getting another chance. Everyone looks young. Even the old people have the smiles of children, as if they have found their smiles, have just learned how to use them after a lifetime of dominion and terror. You can look out at those rich black horse-plowed fields, carved out at the edges of the wilderness, and feel grace and hope and peace and promise coming up out of the fresh-cut soil.

I came to study bears. There are a lot of bears in Romania—big bears, like our Alaskan brown bears, or grizzlies. Where once our country had over one hundred thousand grizzlies, we now have only a few hundred in all of the lower forty-eight states. Romania, on the other hand—the entire country is only two-thirds the size of Montana—has, it is estimated, between six thousand and eight thousand brown bears, *Ursus arctos,* the European (and Asian and Alaskan) equivalent of *Ursus horribilis,* our great grizzly bear.

The reason Romania has so many bears is because Ceausescu liked to kill them—because he was a megalomaniac and had one hundred thousand secret police to carry out his wishes, because it thrilled him to drop dead in their tracks big, magnificent wild and free creatures that were infinitely more powerful than he. Ceausescu outlawed the hunting of bears. He saw to it that there was an aggressive breeding program in zoos and released hundreds of bears into the wild. He didn't want them to be hard to find. He was a small man, plump, out-of-shape man, and later an old man (and then a dead man). He would prowl the mountain roads in his limousine, looking for anything to shoot, but especially bears. He had his Forest Guards set up bait stations to draw the bears into feeding troughs, and he had a cabin—a shooting gallery—built twenty-five yards away from the troughs so he could shoot these semitamed beasts who came in from the woods to feed. He would shoot eight or nine in a day in this manner; but because he was the only one doing it and because he was so aggressive in his feeding and breeding programs, the bear population in Romania leapt from about 860 animals in 1959 to its present number.

And that is this story. I came in naive and left feeling guilty. What was I supposed to say—"Nicolae Ceausescu was a very bad man. He was cruel to bears"? Wouldn't it be a slap in the face to all the lost stories of the butchered?

I still don't quite know how to answer this guilt. I went up into the mountains of Transylvania and looked hard at the bears. In the end, I think this focus had less to do with loving bears and the wilderness and more to do with averting my eyes—if not my heart—from everything else.

A friend of mine from Alaska, Sue Farnham, had been a Fulbright Scholar in Romania the year before, in the city of Sibiu. She had told me some about Ceausescu's legacy of terror and about his legacy with bears. I'd been gathering the scientific papers published by Romania's sole bear biologist, Peter Weber, formerly from West Germany but now living in the town of Medias, in Romania. Weber's project area was in Transylvania, where Ceausescu had done most of his killing, just outside the village of Varsag, where these bear-feeding stations were.

The area was already ideal habitat for bears—dark, heavy spruce forests for hiding cover; small, open meadows along fast rivers; and giant beech forests that produced butter-rich mast for the bears before their hibernation from December to March. Aided by Ceausescu's feeding troughs, the bears flourished even further, using this central "honeyhole" as a core area from which to expand their population. Ceausescu came to power in 1965, riding a wave of jingoistic chip-on-the-shoulder nationalist sentiment that promoted aggrandized feelings of self-importance among urban Romanians, which led to a nationalist fervor not at all dissimilar to that going on in the Balkans. In this country, he might have been called a populist, manipulating the populace's emotions through nationalistic little growls and platitudes, not unlike those of Ross Perot. It goes without saying that our government supported him because he was a "maverick," a thorn in everyone's side.

So all these bears were holed up in this one fairly small mountain range in the north-central part of the state. A lot of them were getting shot by His Nibs, but those that weren't were going into hibernation fatter and healthier each year, were having more cubs, with lower cub mortality rates—and because the biggest bears tended to dominate the feeding troughs, a lot of the smaller and medium-sized bears were displaced. The Carpathians swelled with bears, and then the bears began to spill out of the mountains into the flatlands, some of them traveling great distances. Bears were beginning to show up in places where they hadn't been seen in centuries. In the Carpathians, around that high mountain village of Varsag, they were beginning to cause increasing difficulty for the shepherds and their small herds of sheep.

Well, actually that's the chickenshit bureaucratic kind of language that the federal land abusers and thieves use in this country, such as the timber companies, and so it's not fair to attack them on the one hand and then use their language of deceit and obfuscation, on the other hand—when it suits me. I owe it to the bear and the wild landscape to say it straight out: Some of those bears were eating a *lot* of sheep, and every now and then something would go wrong and they'd kill a person, too. I'm not saying they'd kill eighty thousand people, as Ceausescu's secret police did, or five hundred thousand Native Americans, as our government did—but they'd knock off a human about every third year or so.

After Ceausescu was overthrown and executed in 1989, reports about this extraordinary bear density (in some places, almost one bear per square kilometer) began to leak out of the country. He'd kept all the scientific reports suppressed, perhaps because he feared they'd dilute the impression he had of himself as the great hunter. Ceausescu's name occupies ninety-seven of the first one hundred entries in the record books for trophy bears killed in Romania, as well as throughout much of the rest of Europe and in the Ukraine—and it must be assumed it would have deflated his self-image, perhaps even delisted him from the record books, if the world knew that all of his trophies were, between sips of whiskey, shot from a cabin window twenty-five yards away from a feeding trough and that some of the bears were such regular and frequent visitors that Ceausescu had names for each of them. No, it wouldn't do.

The good news was that if he was killing bears, he wasn't killing people. The bad news of course was—beyond the destruction of sentient beings—that his thirst for blood, violence, and madness was being stoked. He didn't just shoot bears. He shot wild boars, stags, deer—anything that moved, he blew it away.

I like very much the notion that the bears, like the Romanians, have outlasted Ceausescu and that having done so, they are now prospering while he is part of that soil. Whether his vaporous ethers will rise back up, as has occurred again and again in the past (in the seventeenth century, thirteen thousand Romanians were killed in three weeks by their own government), or whether the glowing forces of lights will finally gain the upper hand remains to be seen. That is the dance of humanity as well as the dance of nature, of the universe—the dark and the light. This spring, with those incredible cherry

Laszlo Nagy, Sowing Seeds, Romania, 1993

blossoms and the smiles of raw new capitalism, it seems like maybe the dark blood will stay down in the soil just a little while longer, at least.

When I saw a newspaper article in this country in 1992 titled "Romania Giving Away Bears," I knew I had to go. Nowhere in the rest of the article was there discussion of a giveaway program, but that didn't matter; in my eco-fairy mind all sorts of romantic, quixotic nonsense began to take place. I pictured walking through the woods and happening upon three young cubs playing beneath a beech tree while their mother was off napping or raiding a village. I imagined picking up a couple of cubs—a male and a female—and fitting them into a carry-on duffel bag in a tangle of sweaters and jeans. They would be still and quiet, like hooded birds. If noticed in the x-ray machine as being alive and not stuffed, I would simply explain they were my pets. What? They couldn't go on the plane with me? Oh, all right, I'd buy one of those little portable kennels for them. Sorry, I didn't know that I couldn't bring them with me.

Actually, I hadn't brought a duffel bag. I'd brought my old Kelty backpack and tent, sleeping bag, cookware, and day pack. Sue, who was to be my guide in the country, had a friend in Sibiu named Joachim who could speak Russian and Romanian. The biologist Peter Weber had answered one of my letters several months earlier in hard, broken English, saying in effect that he'd be happy to show me some bears, that it would just be a matter of time.

"If I obtain more information about your work and intentions, I can try to go for a short time with you in my better bear-territories. Perhaps—it depends on your intentions, and not at least from my time also. . . . If you have enough time, it is nearly sure that you can meet some bears there."

I got the impression Weber was a little chafed that all I could speak was English. He asked at the end of his letter, somewhat testily I thought, "Please excuse my faulty English; my German, Hungarian, and Romanian is much better. A question, in what language will you discuss with the people from bear-land?"

I couldn't very well tell Weber that my intentions were centered around smuggling bears. I wasn't even confident of representing myself as an international broker of bears, visiting on behalf of the American and Mexican governments. Instead, I wrote back and told Weber an also-truth—that I admired the way people in Europe were able to coexist with the great bears, rather than killing every last one of them in a spasmodic twitching frenzy of neuroticism, as was done in this country, where there were far fewer people, far fewer bears, and more "big" country. Europe—and especially Romania—might have some lessons or insights to offer America, I said, in the coming century, as we too begin to age. And that was mostly the truth. I just left out the part about the duffel bag.

A part that I did not leave out, but should have, from my return letter— once I'd tracked Weber down—was that (by the way) I wouldn't be coming alone: that my friend, Sue, would be coming, and of course a photographer, Ray, and perhaps another friend of ours from Alaska, Carolyn (who knew a little German, but not much), and then of course Sue's friend Joachim.

I didn't hear from Weber again; he submerged like an old fish sulking beneath a big log, hidden safely in the faraway depths of a foreign country. He was safe, and he knew it. Frantic phone calls to our man Joachim in Sibiu across the twelve-hour time change—me from the pay phone in Yaak, Montana, and the phone systems in Romania heartless and all but absent. Telegrams to Weber, unanswered, and to Joachim, who was in Germany and had heard that Weber was off in the Danube Delta, doing bird counts—hiding in the cattails, is how I pictured it, waiting until these Americans' ardor for bears cooled.

The time to go was March or April, when there was still snow in the mountains and when the bears would be hungry and active, having just come out of hibernation. They'd be easy to track in the snow. But March and April slid past in Weber-silence, as had all the many previous months, so in May we just bought our tickets and headed over there. Our friend Carolyn dropped out, perhaps unnerved by the readings her shit detector was giving off, and alas, we got word at the last second that Joachim had other plans. But we did have Ray the photographer and plane tickets. We'd be over there for eight days, which was the longest that I could stand to be away from my one-year-old daughter. The photographer was cool. Usually they are egomaniacal drug addicts, or the ones I always seem to end up working with are, but not this guy. He was tall and quiet, his hair cropped short like a swimmer's, and he had a one-year-old son at home, whom he'd never been away from. We would rent a car and drive north to Transylvania, stop the car at the edge of a forest, and walk into the Carpathian Mountains. We would hike twenty or more miles a day, looking for bear sign—fresh grassy shit, mostly—and when we found the

Vale, Romania, 1994

fresh sign, we'd hide in the woods at the edge of a high meadow with binocu-
lars each dawn and each dusk and under the full moon.

A man suddenly throws open the double doors and stares round-eyed out at
us. He is short and wide, with a full red beard, red thinning hair, and thick
glasses. He's dressed in Swiss-style mountaineering shorts and boots and a
plaid shirt and is wearing a rucksack; it's got to be Weber.

  "Come in," he says. "I have been expecting you since daylight. We are late,"
he says, stalking around and supervising our pack-stripping. He is especially
stern with Sue, standing over her and repeating his mantra, "We will be gone
only a few days—we will need nothing." We get the hint as he motions to the
undeniable impossibility of his car. The car looks like a thimble; it's hard to
imagine Weber fitting in it, much less four people.

After having left most of our identity behind stuffed in the rental car, we rocket off, bouncing crazily over cobblestones, the car's little engine screaming, frame and underbelly clanging and throwing sparks. We are driving so fast that we're out of Medias before we know it, barreling down one of those long, forever straight roads that runs through Romanian farmland, much like the roads in North Texas and the Midwest. Weber blows past wobbly cyclists on their old black bikes, brushes at over 110 kilometers per hour the flanks of horses pulling carts. It's like being in a go-cart. We're too terrified to say anything and too timid, due to our lateness and Weber's generosity in taking us to his bear haunts, to protest.

"Bears are going into the woollies," Weber explains, pointing to the fields below us—our road is climbing higher into the foothills. At first I think he is talking about sheep, *woollies,* but then understand that he is saying *valleys*— that the bears are coming out of the mountains and into the valleys. Five or six bears, he says, have come out into the big maize fields just north and east of Medias. The Communist party busted up the small farms, Weber says, and cultivated huge fields of monocultures—grains—that happened to be a food the bears could eat and even relished. As I understand Weber to tell it, the bears grew bold because the farmers had no guns, and the bears would be out grazing in the fields even in daytime: much as grizzlies must have done along our sun country's Rocky Mountain Front in the early 1800s and for all the centuries before Lewis and Clark came through.

It sounds like the bears went where they pleased (as long as Ceausescu wasn't around); they chased farmers away from their horse-and-plows, away from the occasional tractors, and just ate that grain and got bigger and fatter and more numerous. At least something was prospering, besides Ceausescu and his wife.

We turn down out of the foothills, head across the valley floor—going north again—and are soon rocketing through ancient villages in which we see no other cars; nothing but horses. Sometimes Weber uses the horn, but only when a collision seems certain; if it's going to be close, he just mashes down harder on the accelerator, swerving into the paths of oncoming horsecarts and then back into our lane.

Scientists like Weber are rediscovering, and reproving, an old truth that every ancient culture of man since the dawn of time has known (except for those from the period 1800 through 1900, the Industrial Revolution, who lost sight of this truth, let it slip through the cracks, like so much other knowl-edge): *Bears are intelligent. Bears have feelings.*

I know that Weber knows this. When you're out in the woods with the ani-mal, living in the animal's environment—its community—you can't help but see and know these things.

But then, when you come back to Parliament, or when you travel back to your office and sit at your desk to write your scientific papers, how hard it must be to leave those things, those feelings you've felt in the woods, out and to lean on the dry ways of the scientist and the scientific methods that are so necessary to achieve dialogue and respect with your peers.

One of my goals on this trip will be to see if I can get Weber to open up and admit that he feels the bear has a special force, or spirit, which attracted him in the first place to the creature he has chosen to study and whom he re-spects and admires.

Every time I edge near the question, however, Weber shies away, almost vehemently.

I know he believes what I'm asking. It's just got to be that his whole iden-tity—the cold, aloof scientist—is at stake. It's how he was taught and how he's conducted his work, both under the Communists and now under Inde-pendence. A foreigner can't come over and in a week coax a life-held secret from a man— especially not a secret that's been guarded under such an iron regime, where the cold exacting ways of hard science fit the communist theory so safely. I won't blame Weber if I can't get him to admit even a simple truth such as: Yes, he has a fondness for the bear—but I am going to stay after him about it. I think it's what his peers in the rest of the world are starting to un-derstand—that it's okay after all to admit an attachment, even a passion for your subject—and I want Weber to at least consider this. I want to plant that idea in his mind, to rest there in the coming years.

Still, he bolts whenever I move in that direction. When I ask him if the vil-lagers care for the bear, if they feel it's special, if they'd be unhappy if their bear went extinct, Weber replies sharply, "That is not the point. I believe that they accept the bear. They arrange their lives with the bear."

What it sounds like is that bears and humans are living together. The pic-ture that is emerging is of two slightly separate nations—the bears, and the vil-lagers—each wary of the other, but by and large too busy with their own lives to engage in war: a relationship not without skirmishes. The picture I am beginning to get is of the bears hanging out in the dark woods, watching, making almost a game of it: rushing down out of the mountains and nailing an occasional cow when they absolutely can't resist it, or—if a shepherd falls asleep on the job or is careless—moving through a flock of sheep.

Once in a blue moon, attacking a villager, when still more rules are ignored.

Something some bear experts have been saying in the States for a long time, and that other biologists are just starting to acknowledge, is that bears, even grizzlies, are more social than people give them credit for. Because they didn't see bears hanging out together a lot, the biologists used to assume they were solitary. They didn't stop to think that it was more a function of food availability, that a grizzly's a big animal and needs a lot of space just to take care of its nutritional requirements. Back when there was more habitat, there were reports of several bears together, even out on the plains, but because those reports were anecdotal, they were not properly filed, and because they were in the past, they were largely ignored.

In Romania, however, Weber (with his three thousand hours logged in the captain's chair) is seeing rich social behavior, given the stability and extravagance of the bears' food source—their external feeding station.

"They *are* social," Weber says. "I numbered sometimes at one place in late afternoon and evening twenty-eight bears."

How much easier for us to trash their habitat and treat them poorly, like second-class citizens, I think, if we make the convenient assumption that they are passionless loners, all of them, and really won't miss each other when they're extinct.

Weber agrees—sort of. (I don't lay any of that communist/capitalist stuff on him.) "It is a way that bear is described to hunters or from hunters from the first part of the century," he says—this stereotype of the grizzly as being unfeeling and antisocial. "That the bear is an individual who walks always alone and is bad, is dangerous—an animal whom if one bear smells another bear, the fight is on. And the big bear kills the little bear, and-and-and," Weber says—his expression for "*et cetera.*"

"But much more time," Weber says, "I see bears staying together, or not far from one another."

We pass through another village—the villagers seem to know Weber's car, as they scatter the second they see it coming—and then we climb a hill north of town, cross a dam, and are on a dirt road, raising dust, driving up a canyon. White blossoms—dogwood?—float in the woods on the hillsides above us, and I can feel that we are closer to the bears.

"We are now in the East Carpathians," Weber says, with pleasure, almost relief—does he feel it too, down in the cities, that sense of entrapment—and he slows from one hundred to eighty.

We follow the rushing little creek, higher into the mountains, which are a lush and lovely mix of the dark firs and the white-barked birch, their pale light green leaves glowing with the newness of spring, and the diversity of the oak-hornbeam forest mixed with that. It's like the best of the northern Rockies mixed with the best of New England—that's the only way I know how to describe it. And then, as I'm thinking that, we pass through long open stretches of meadow, with the sky so blue that I am reminded of the high villages in New Mexico—sycamores, or their equivalent along the stream, giant cottonwoods, and the smell of woodsmoke. Weber's relaxed further, slowing from eighty to sixty, as he rolls his window down and sniffs the clean spring air.

We pass an old stone castle, and when I ask its name—something that sounds like "fruit-fisher"—Weber frowns and says, "You mustn't try to write it. It's too heavy for American languages."

Fine, fine. We pass woodcutters, chopping bundles of wood, little piss-ant switch sizes of wood, stacking them on wood ricks and then bundling them with wire, to sell for firewood. They're using axes, of course, not chain saws. I like the sound their axes make against the wood, and the absence of the saw's gurgle-and-roar. I'm not sure what wood they're cutting—if they can only cut blown-down branches, or what. Weber said that any tree that's cut from the forest has to be bought, that the Forest Guard comes out with the purchaser and they mark that tree after it's been paid for. The trees exist for the people's use, not the corporations, because there aren't any corporations. I guess all that will change soon enough.

I must remember to send Weber some pictures of what our own "forest guards," the U.S. Forest Service, have done—how they've spent taxpayers' money to build roads through and across delicate streams and up, over, and around wild mountains so the private corporations such as Champion (cut-and-run) International can cruise in, scrape the hills bare (cutting not one certain tree but two hundred acres of every standing living fern snag bush tree blade of grass), building a mill there, becoming the central core of the community, and then abandoning the community when all the trees have been cut (but

Man Cutting Wood, Varsag, Romania, 1993

first dividing the land into the smallest permissible fragments and selling it off to real estate investors).

We wind up into the mountains. It's nice to hear the occasional ax ring out. Weber told us earlier that with the continuing increases in food prices, there are getting to be more and more wood thieves, people who fell a tree at night then try (by traveling only at night) to sneak it into a city and sell it there.

Weber says that the villagers don't like the wood thieves and that they are "looking out" for the thieves who try and cut the trees, and again I have to marvel at the difference in scale. I wonder how these villagers who get all hot and bothered about one missing tree would feel about being gone from their forest a week, as I've experienced, and coming back to find the whole hillside gone, the whole side of the mountain—gone.

It's all new, this country, and I'm devouring it with my eyes. It must be like Montana in the forties, I think, and maybe even farther back than that: maybe the thirties, or twenties. It's delicious. I didn't think I'd ever see such a thing. I thought I'd been born too late.

We turn up one of the little creases in the hills—the north slopes dense and dark with timber, and the south slopes open with patches of meadow—and follow a creek up to the cabin where we'll be staying. One of the last things Ceausescu did was bring electricity into Varsag by damming one of the big rivers and then marching the power lines' grid up through the mountains to his hunting village. He flooded an entire village system so he could still use his hair dryer when he came to the mountains. Maybe he used his electric chain saw to butcher the bears after he'd shot them. (Although by all accounts, the only time he got close to them was to have his picture taken. A source in Bucharest, whom I don't wish to identify, writes, "He loved to be filmed in front of long lines of beasts and trophies"—almost, it sounds, as if to balance his raging wild evil against that which he believed the "beasts" represent.)

I'm glad electricity has come to Varsag. (The cabin where we'll be staying hasn't been hooked up yet.) The mounds of earth where they dug and set the giant concrete pillars are still fresh and disturbed. But this region has been in the dark for all of the history of man—the bitter border disputes with Hungary and Russia, the Magyars, the Saxons—and the dark hills are rich with the blood and suffering, the constant hardship. No matter that the project may have cost tens of millions of dollars to come all the way to a town of twelve

Peter Boldizsar & Dogs, Vale, Romania, 1994

hundred, and no matter that Ceausescu had in mind to dam every flowing river to industrialize his country, to turn his back on its natural resources, to try—with his savage inferiority complex—to deny his country's rural identity.

Weber says his favorite bird of all is in this forest, the black stork, and that we might see it.

"Is there anything we should know?" I ask him as we drive up to our cabin. Weber's as excited as a kid.

"The dogs," he says, talking about the sheepherders' dogs, the ones with the savage spiked collars—the hundred-pound brutes with sticks dangling from their collars to keep them from running away after stags or wolves. (The sticks flip and bounce against the dogs' chests, bounce up and hit them in the chin, and make long running undesirable.) "If any dogs charge you, sit still to keep them from getting you," Weber cautions us. "That is their job, to catch you, not kill you." Just sit still and let them surround you, he says, and let them bark at you, and you'll probably be okay.

"Come on, follow me!" Weber cries, running up the steep hill to the cabin. He's like a colt—why shouldn't he be? All of the Harghita Mountains (what this particular chain of the Carpathians is called) are his. It's spring, and communism is gone!

Say it, Peter—*you love the bear,* right?

The cabin's already stocked with food, candles, pots and pans, and such. We carry our sleeping bags in—we have no luggage, save our day packs with our toothbrushes and canteens in them—and then Weber drives up to a place where we can look for bears, about half an hour out of Varsag, up into the hills above the town. We drive to the end of a new logging road and park at the end of the road, below a hillside of alder and willow, in the vee of a canyon: dark woods on one slope, above a little creek that comes out of the low mountains, and a large and open slope of alder and willow on the other side. What it is, is a big-ass clearcut from not so long ago, and there are little hunting platforms all along its edges. Weber explains that mostly stags—the European red deer, close cousin to our elk—are hunted here, but that back in the forests there may be some bears. The trees have been cleared off that one slope, he explains, so the hunters can see to shoot. It's too big and open, and if it were smaller, bears (and elk) would be more comfortable and would use it more often. I want to tell Weber about the distance-to-cover ratios that need to be considered when "creating" a mosaic of habitats for bear management. I want to tell him, too, that the best habitat is no roads at all, just total security, but then, I remember that actually the best kind of habitat, from a strictly nutritional standpoint, is to have a zookeeper who brings you a dead horse every other day.

I try to make myself understood, and Weber nods, points to the clear-cut hillside, which is drying out without its protective canopy and is stubby and barren, and agrees. "The bears don't go with pleasure through such places," he says. They go around it and stick to the dark forests. This cut-over hill is mostly for shooting elk. It's so dry and stump-stubbled that I wouldn't think the elk would go with pleasure through such a place either, but then I remember what my acquaintance in Bucharest said about the hunting, and the suspicion comes to me that the elk are driven across the big opening.

"They say that once, he [Ceausescu]," wrote my contact, "sat on a small chair in the cabin of cable railway, which was stopped on purpose exactly above the valley at the lowest height, and they chased the animals directly in front of his gun!"

The part of the mountain that hasn't been cut looks perfect for bears, as well as elk. We walk along the creek in the suddenly hot sun (the heat bouncing back at us). Weber is relaxing further still; walking slowly through the ferns, sauntering. He points out a barely noticeable depression in the ferns and tells us that "the bear had made his way there." I don't know whether he's bullshitting us or not.

Weber points out some hoof marks in the muddy creek bank—the red deer, the stag, just like our elk, or wapiti. Weber chuckles and then makes fun of the high-pitched bugle that the American elk makes, during the rut. The Romanian red deer makes a deep bull-like grunt, very majestic, but Weber says he about fell out of his chair laughing the time he saw the film of the American elk bugling.

"It was such a little sound from such a big animal!" he cries. "Trying so hard to make the noise, but all that came out was this little squeak!"

I don't know what to say. I'd always thought the high, reedy bugle of elk in the autumn was one of the loveliest sounds in the mountains. How strange it is to realize that someone else finds it amusing, even laughable!

(Later I will read that the reason our elk have developed the high squealing bugle is that they evolved out on the plains, where high sounds carry better.

Only recently have we chased our elk up into the mountains. The red deer, however, have always lived in the mountains, and in deep woods, a deep sound will travel slower but farther—the deep sound waves will kind of bend around trees and creep over fallen logs, rather than ricocheting and deflecting the way shrill sound waves would. But I don't know any of this at the time, so I just have to suffer Weber's teasing in silence as well as his jokes about the American elk not being very manly.)

We cross the creek and wander down an old shady logging road beneath giant cathedrallike firs. I like the old forest. But it ends soon enough—too soon—and we find ourselves bushwhacking across another stubble-stumped razor-slash clear-cut. Weber seem halfway proud of the clearing in the forest—the whole side of one foothill—and explains that it was necessary for the hunters in their boxes (he points to various blinds stationed around the edges of the clear-cut) to be able to see in order to shoot.

The clear-cut's too big for bears; this area is for hunting, or shooting, I should say, the red deer. When I tell Weber that it doesn't look like what we'd call good bear habitat back in the States, he agrees. "The bears don't go with pleasure through such places," he says.

We leave the clear-cut and walk down a foot trail back into the woods. At nearly every bend, Weber has a story to tell—how a bear was taken here, or there, or here . . .

"Where are the bears, Peter?" Ray asks. "I want to see bears." Ray points to his camera. "If I don't get pictures of bears, I won't be paid. My children will go hungry. Where are the bears, Peter?"

Weber smiles a small smile. I'd like to know what he is thinking. He studies Ray for maybe twenty seconds with that smile. Maybe what he's thinking is, *What an odd bird this American is; what a long distance to travel, just to give sass* . . .

"We will try and find you a bear," Weber says. He points down a lane, a wide trail. "You go down there and wait. We will go up ahead and then try to drive the bear out of the bushes toward you. When he runs out, you must take his picture."

He smiles at Ray a bit longer, enough to make Ray wonder if that's the kind of look he gives to people whom he'll never see again, and then shoos him down the lane. Ray looks back at us with a "Will I ever see you guys again?" look of his own, only unlike Weber, he is not smiling. Weber and Sue and I troop off up to the hill, into the forest. Weber stops and points to a pretty big bear scat in the middle of the trail. Pretty fresh. Weber stands over it for a minute but does not pick it up and examine it like the bear biologists back in the States do. Then we move on, across a marshy spring and through such perfect grizzly habitat that I have to wonder if it's not a joke, if maybe Weber knows of some old giant grizzly holed up here above this seep and really does intend to run the bear right over the top of Ray.

Instead, Weber sits down under another tree. In his little knit mountaineer's cap and his long-sleeve wool shirt and his hiking shorts with their suspenders and his high stockings with their garters, he looks totally Bavarian: as if he plans to hum some little song, then fall asleep with a half-eaten piece of cheese in his hand and an empty beer bottle by his side.

Instead, what he does is pull out a thin metal whistle and blow on it, as if it were a flute.

Too much, I think, with a sinking feeling. *He's got these bears so conditioned to people and to feeding that he whistles for them!*

I sit there next to him, in almost unbearable agony, as we wait for the great bear to appear, shamefaced and stripped of dignity, begging for a handout from the Lord of the Universe, man.

Finally I have to ask Weber about the whistle, and he laughs.

"Oh, no," he says. "This imitates the sound of a ground squirrel, which the bear loves to eat."

But none presents itself—nor should it, with the gang of us sitting there and talking and filling the air with our life-scent. (A grizzly can smell a dead elk carcass at a distance of up to seven miles.)

We walk back to where Ray's sitting with his camera. His eyes widen when he hears our footfalls crunching twigs—when he first looks up at us. He will not be cocky—will not tease Peter—for a good thirty minutes after that.

Enough about bears.

We've seen one scat; they're here. It's time to go back into the village now. Like any good bear biologist, Weber understands that in today's world, how people get along with the bears is at least half the battle to keep the bear around. In the not-so-old days—two, three years ago—in Romania, if you shot one of "the dictator's" bears, you'd be sure to end up in jail, or worse. This was pretty good prevention against illegal bear killings (except by the dictator himself). I'm curious to see how people feel about the bears, now that

Transylvanian Brown Bear, Varsag, Romania, 1994

Ceausescu is gone, curious to see if they were just pretending to get along with them out of necessity, or if they really feel that way now that they're free to speak out. The Montana grizzly biologist Dr. Chuck Jonkel has said that the three vital requirements for grizzly bears are "habitat, habitat, habitat," and he's exactly right, of course; and one of those components of habitat is security from human beings who will kill them. In a country like the United States, where people kill grizzlies simply out of habit or culture or unthinking hatefulness, that security must take the form of large roadless areas set aside to remain roadless through generation after generation of bears—forever, if that's how long you wish to have grizzlies in your country.

But in Romania, the downtown main dirt road of Varsag can be security habitat, if no one harasses the bear. (There was that one farmer who ran at the grizzly with his pitchfork, but that was an anomaly.)

That's where Weber's investing his time. The food and cover is already present. And the movements of bears are not too terribly complex: Like any of us, they're pretty regular. They eat, they shit, they sleep; they wander around. Every great now and then—like any of us—they may do something surprising. They sleep through most of the winter. So there's that aspect of the biology of it. But the people who live at the edge of the dark woods and rolling mountains—the people who have carved out a little opening in the woods and who have lived there at sustainable levels for the last several centuries—their relationship enters into the biology of it just as surely as do the other factors: the winds and the snows, the weather, the lush green springs, the red deer, the ferns, the mushrooms, the oak-hornbeam crop of each year . . .

And so that's where Weber spends a fair amount of his time: stroking the locals. Shooting the shit with the old Magyars, the terrible Magyars. Hearing the latest about which bear ate which sheep, where, and how many. Smiling, just hanging out, and being calm, steadfast, and casual in the swarming heart of bear country.

Are they swarming? I had kind of half hoped, if not expected, to see more sign of them, to find a big muddy footprint in the center of the road through towns. To be truthful, that was one pretty small, dried-out pitiful-looking little turd we came across back down by the marsh. What if it is all some communist propaganda scam; what if they don't have any more bears than the rest of us?

Epilogue

Before it was over, we ended up seeing bears everywhere: bears still coming into the feeding troughs, bears running through the woods, bears scampering across the roads at night. We drank *palinka* in the villages with people who lived arm-in-arm with the bears and who seemed still giddy with the palpable combination of springtime and freedom: another springtime, one of the first in which the dictator was truly gone.

It's hard to imagine, given the country's history, that trouble won't return. But this spring, amid the apple and cherry and plum blossoms, it seems a long way off. In many ways, the villagers in Transylvania seem more secure—on more stable ground—than do our own country's villagers. The land still seems healthy, here, and it seems as if that is one of the things that is willing to give them a second, or third, or fourth, or twentieth chance.

I wish we still had as many grizzlies as Romania. In a way that I find myself unable to prove or even properly articulate, the grizzlies and their wild country seem like the perfect antidote to the ways of dictators.

silver fragments of broken mirrors

Mariana Kawall Leal Ferreira

> Tell me about the mirror. We always thought you were different because you never looked in the mirror. My wife always used to say, "Mariana's house doesn't have a mirror." Look at yourself in that piece of glass I gave you. Why do you insist on being an Indian? I can't tell you who I am if you don't tell me who you are. Tell me who you are. Look in the mirror, and tell me who you are.

I want to listen, but I have to speak. Sabino's words suspend my certainties, consuming mirages of a past I long to keep hidden. Sabino insists on asking me about disentanglements, turning points, historical scars. The cracked, multiple angles of the broken mirror I hold between my fingers tell me nothing about myself. Better still, they say too much.

During my visit to the Xingu Indian Park in 1990, where most of the Kayabi population now live, Sabino asked me to write down his life history so that "the children can read in school [at the Tuiararé Village[1]] the history of the Kayabi, and the white men can learn what it means to be Kayabi today." Narrated without interruption for eight days in the Kayabi language, the history was preceded by our conversation about his current status as a *uriat*. After having his soul abducted by *añang,* a malignant spirit, Sabino became partially paralyzed—"finally able to rest," in his own words.

His current situation as a *uriat* conflicts with the signs of disease and disability that health professionals in Brasília, the country's capital, had warned me of. He was now, according to them, "severely impaired," a "useless old man" who had recently suffered a stroke. To my surprise, and in response to my concerns about how he felt and how he managed to get along with his left arm and leg paralyzed, Sabino said he had never been so relaxed and at ease in his whole life.

> You also think I am diseased, don't you? Doctors keep telling me how diseased I am, giving me medicine, telling me to exercise. But you just don't understand; you don't understand because you know nothing about me. If you only knew how much I suffered my whole life since my mother's death when I was four years old; my life on the rubber plantations; working for FUNAI [the National Indian Foundation] or its predecessor, SPI, on attraction fronts and everything else, then you would understand how I feel. Now I can finally rest, look after the Kayabi people in my dreams, talk to them, advise them, tell them stories, and sing to them. I don't need to walk or run, do hard work in the gardens, hunt or fish, construct houses or canoes. See the young men working outside? They are almost done building houses for this big village I always dreamed of. Now listen to me, just listen.

beyond the marvels of modernity

Sabino's gaze turns backward, but there is no return. Spaces void of love and dreams. Promises that never become: the return to a territory the Kayabi abandoned in the late fifties in exchange for beads, cups of coffee, firearms, and antibiotics.[2] The Kayabi geography has been overtaken by *añang,* malignant spirits that insist on probing into the celestial realm of the cosmos. *Añang* claim the fame for deeds the Kayabi attribute to Tuiararé, the Creator of mankind. Sabino's eye explores a no-man's-land, identifying perverse beings that threaten the Kayabi universe.

> The *añang* are all over the place. They are mean, mean just like the white men who have made our people suffer so much. Actually, I think the *añang* are white men's spirits since they are the meanest beings. The white men kill us with their firearms and deadly illnesses. The *añang* kill us with their *mamaévévé* [magical objects] and their also deadly illnesses. All the same, huh? I saw *añang* in many different occasions when I

worked on rubber plantations, but they were never able to capture my soul. All they did was make me ill, but I soon got better. Only last year was *añang* able to steal my soul. But luckily I recovered it.

Kidnapped by *añang*,[3] Sabino's soul wanders aimlessly, beyond terrestrial frontiers, into the depths of time and space through the different domains of the Kayabi cosmos. In flight, the *uriat* communicates with different spirits, animals, animated objects, and people, Kayabi or not. The *uriat*'s extraordinary capacity to communicate with such different beings through songs, discourse, or dreams is what grants him so much respect from his own people. This mystical or cosmic voyage is a product of the abduction of his soul. It reveals, in dreams or in trances, his own symbolic death and resurrection. Sabino departs from a more immediate level of reality into a higher state of consciousness. The shaman, in ecstasy, gradually becomes acquainted with the spiritual realm and learns different chants, the core of several Kayabi healing therapies.

In the early thirties, Sabino and his mother, with other Kayabi born in villages spread along the Teles Pires River in the state of Mato Grosso, Central Brazil, moved to the Pedro Dantas Indian Post. This was the first administrative unit set up by the former Indian Protection Service (SPI) to "pacify" the isolated Kayabi. While some groups remained hostile to such civilizing attempts—either attacking "pacifying fronts," missionary posts, rubber camps, and prospecting sites or else moving away from pioneers who had recently reached their territories—others sought employment, health care, or refuge at these settlements. Employment, health care, and refuge have meant slavery, sickness, and death to native Brazilian peoples for centuries, as it did to Sabino and his close kin.

> The merchandise brought to the Pedro Dantas Post by Inário [an SPI employee] was contaminated with measles. All the Kayabi fell sick. First, ten of them, and then my mother. None of the white men got sick. The health aide, Antonio Pretenso [also hired by SPI], did not take good care of us. He administered "snake medication" [antivenom serum] to the dying Kayabi, to help kill them faster. As soon as he gave people the shot, they would die. And this is how this guy helped the measles kill the Kayabi people. . . . In two weeks, 198 people died; only 40 survived.

fictitious bird tracks unleashing the imaginary

Sabino and I retreat into dialogue; he asks me how I feel. My tears drop onto the dusty floor of his thatched-roof house, weaving words of dismay and stupefaction through my meaningless scribbles. "You write like a bird swiftly running on a beach," says Matareiyup, the *uriat*'s son. The revealing gesture is yet to come: cracked images of a small round mirror Sabino gracefully places on my lap.

"You need a mirror," he tells me once again, after timeless repetitions of a reflectionless stage I went through in the early eighties. My mirrorless house in the Xingu Indian Park conflicted with the white men's ontology of better knowing the Other through its own opaque reflection on a screen. I shiver and close my eyes. So does the shaman, Sabino, the *uriat*.

In flight, Sabino's body swallows the world. The cosmic journey is one of identification, of naming exquisite beings that inhabit different cosmic realms. By equating *añang* with colonizers, the causes of evil are sinisterly revealed. Sabino intervenes in a once "supernatural" cosmos, made less super and even less natural by providing its interlocutors with names: governmental pacifying agents, experts in Indian affairs, missionaries, rubber tappers, gold prospectors, city marshals, health officials, photographers, anthropologists. The so-called modernizers of an "empty space"—Central Brazil.

A theatrical space is produced. Indians are phantasms that haunt the white men; invisible entities that the colonizer's desire does not want to fix in the picture of a world named "new." The discovery of remote or empty regions of the planet was the founding basis for a perverse geometric relation that pretended to catch, manipulate, and capture the phantasms it refused to see in its field of vision—refused to see because of the very fragmented bodies it made decay, left rotting, moribund states of pitiless anger, the images of death.

the gaze of death: the intractable

Still crying for his mother, Sabino traveled several days by canoe with Uncle Kawaip, one of the survivors of the measles epidemic, to meet Captain Júlio. Sabino's older brother, Júlio, had remained hostile to the white men's "pacifying" attempts until then. Outraged by the news, Júlio set off with Sabino back to the Indian Post to kill the men who had murdered the Kayabi. They were obviously all gone by then, and Captain Júlio became the leader of the Kayabi who had not succumbed. His commitment to avenge the death of his people was struck short by his own death in the next measles epidemic, five years later. The SPI employees arbitrarily named Sabino the chief of the Indian Post, which had its name changed by the SPI to Bezerra Indian Post after some infamous Indian "pacifier."

> I told them I did not want to be the chief, that I was married and had a wife and kids to look after. But they did not care. They told me I would be punished, I would be sent to Campo Grande [now the capital of Mato Grosso do Sul, where the SPI headquarters for the area were located] to work for them if I didn't accept the position.

The cracked mirror falls from my lap; Sabino opens his eyes. Embarrassed, I sweat profusely. The heat is tremendous. He tells me to go bathe in the Xingu River, and I do.

Dripping wet, I make my way back to the low stool he points me to, a headless armadillo hand-carved by Matareiyup. "The Kayabi were once headhunters," he says. "That's what drove white men crazy. Because we chopped off their heads when they tortured, raped, and killed us, they thought we were brutes, animals. So they started treating us like animals. They never understood what head- hunting was all about." The barbarity of the colonists' own social relations was thus reflected back onto Brazilian Indian policy but as imputed to the savages or evil figures they wished to colonize.

Juliana, Sabino's wife, hands me an anamorphic object carefully wrapped in cotton fiber. It is a white man's skull. "Here, hold the dead. Does it frighten you?" It is stuffed with light brown hair. I recognize it as my own, the waist-long, braided hair I once cut, back in 1982 or '83, and that vanished mysteriously from the basket above the fireplace. "It was once your hair. We needed it badly then for a dance, but now you may have it." No refusals accepted; it is a gift.

I turn away from the skull, the hair, the woman. "Where are your children? Did they get to use the *tupai* [sling for carrying children] I taught you to weave?" Juliana asks, while tightly wrapping my knees in cotton.

"You always tie them too tight," I reply, but she does not seem to listen. "Tonight we will dance."

I am afraid they will ask me to carry the skull around. I ask Sabino to go on with his narrative, but he asks me instead what I dreamed of at night, what I usually dream about during the day. Swallowed by fright, the sharp edge of the broken mirror I hold tightly onto cuts my thumb; blood taints the bird's tracks on the sand, the bleached piece of paper that holds my nothingness. Where am I in the story I write? How to convey in words my ineffable dreams that I have never told anyone about, that cannot be spoken about?

"I often dream I am falling down, falling from places I want to escape from," I tell the shaman.

"Do you feel like going back to São Paulo, abandoning us once more like you did back then? Were you frightened, as you are now?" "Why do you do this to me?" I ask. "Why do you insist?"

"Tell me about the mirror," he says, "why you don't like them. I am curious. Your house never had a mirror. Why is that?"

Caught by Sabino's piercing gaze, I am both fascinated and helpless. No use staring away from the cracked mirror that gives me my double, both the *I* I try to escape from, and the *me* I cannot hold within myself. Sabino's glasses reflect my dissociated self; I am deceived by my own reflection, taken into the lure of the shaman's gaze. No use closing my eyes. The gaze is not all about vision anyway. It is about power and control. The *uriat*'s visions make me give up control of myself, like a knife cutting through my flesh.

"Mirrors," I murmur. "Mirrors."

repugnant savages: dangerous bodies

Being the chief of the Bezerra Post actually meant supervising the Kayabi who worked for rubber tappers, attesting to the intimate articulation between the federal government's policy for Indian affairs and the colonizing fronts of Central Brazil. Besides the machetes, hoes, axes, and scythes that Sabino needed to clear trails and paths on rubber plantations, he was also given fabric for the women whose husbands worked as rubber gatherers, and a dagger, a rifle, and "five hundred bullets to keep the situation under control." The Kayabi were often sexually abused by non-Indians at the rubber camps, and Sabino was supposed to prevent conflicts between them.

> SPI also ordered me to contact those Kayabi that had not been pacified yet, that were still savages. They convinced me by saying that they could all be dying of some sort of disease and we had to protect them. But I guess what they really wanted were more Indians to work on the [rub-

ber] plantations. They knew how hard we worked and that we also knew our way around pretty well in the jungle. So my first real task was to attract the isolated Kayabi by giving them mirrors, fabric, fishing hooks, machetes, and other tools and promising them that they would soon have much more of whatever they needed or wanted. More and more mirrors were yet to come, see?

untying knots of imaginary servitude

In flight, the shaman is blinded by the sun reflecting on silver-mirrored surfaces: aluminum-roofed houses, mercury-saturated rivers, metal-glowing airplanes. Farther inland, hanging from twisted tropical branches, cheap-framed mirrors, strategically placed by Indian "pacifiers" to unleash the imaginary of the child they want to tame. Zenith of the modern man, mirrors create fictions, produce knowledge. What is the fiction that creates the Indian in the mirror?

"I have always wondered why it is that white men are attracted to mirrors, and why it is that they make us attracted to them. What do you think?" asks Sabino, perceiving my perturbation with the cracked piece of reflecting glass he has given me and that I do not know where to place.

I must give up my fairy tales, hidden secrets I never thought of sharing with anyone. Sabino's speech, as any other, calls for a reply. My silence, tears, sighs, facial expressions, and gestures are not enough. The *uriat* wants me to ascribe meanings to my self with spoken words, to create something, someone, within a world of words.

My flight is one of marvels, of mirrored images coming to reality in flashes of line and light. Ruby reminiscences of my first encounter in 1978 with a Xavante Indian, Mario Juruna, in the city of Barra do Garças, Central Brazil. My nineteen years of age seemed all of a sudden to be enveloped by clouds of red dust that tainted bodies, images, and thoughts during the prolonged Central Brazilian drought season. At the hotel where I was to wait for transportation to the Kuluene Indian Reservation, the manager points to a hammock in the corner of the cafeteria: "No vacancies."

"Who are you?" asks Mario, a Xavante well known in Brazil during the seventies for carrying a tape recorder around to "record the white men's words so that they can't lie all the time."

"Mariana."

"So you've come," he replies, shaving a scarce beard off his round bold face. Mario turns the mirror he uses toward me, while I stare at the tokens of modernity he has chosen to wear: a digital watch, a silver-coated necklace, and a golden ring.

"Did you expect to see me naked?"

"This is you in the mirror. You're another one of them." As I wipe the red dust off my sunburned face, I see myself in the eyes of the Other. "Your eyes are green. Tell me where you come from." No use staring away from my own reflection; the image I do not want to assume is reflected back onto me by Mario's expanding and contracting pupils. The bright sun blinds my field of vision. I am nowhere to be found.

"Is that when you decided to become one of us?" asks Sabino, referring to the transformation that took place in Barra do Garças, when I tried to assume a particular image, that of an Indian. The drama of the mirror-stage played out in our first encounter gave rise to a succession of fantasies that extended far beyond body images to a form of their totalities. Fantasies of an *I* that could never be formed in its totality: "You will never be one of us." The knot of imaginary servitude was nevertheless tied between my own ego and those of the spectators I needed to put onstage. Mirrors, however, put *me* onstage, fulfilling my first object of desire to be recognized by the Other I longed to be.

Back at the rubber gatherers' main settlement, after "successfully" having "pacified" Kayabi Indians from several different villages along the Teles Pires River, Sabino was told by Akamá, an elderly Kayabi woman who cooked for the men at the camp, that several Kayabi women had been raped during his absence. Ready to kill the aggressors, Sabino was calmed down by his "boss," Antonio Bernardino, who promised to "take care" of it the following day. And so he did.

> Bernardino asked me to join the three men that had been found guilty of abusing the women and himself on a small trip in his Toyota truck. We stopped at a clearance, where he made the men dig their own graves, and I myself had to guide them to the very edge of the graves only to see them make the sign of the cross, be shot one by one by Bernardino, and fall dead into their tombs.

"Do you believe in God?" asks Sabino's wife, revealing a purple plastic crucifix inside her bra. "This is the God we have been told over and over to believe in." My flight is one of anger, fostered by the pitiless conversation I have helped to constitute. "Do you believe in God?" she insists.

"You have no God, you have nobody. Is that why you want to become one of us?" says Sabino, showing me a picture of the Kayabi gathered around a flagpole on Independence Day.

"This is your God," says Juliana. "Brazil. Here, you can make a dress out of this for you."

The moth-eaten, foul-smelling yellow and green Brazilian banner she hands me weaves images of a grotesque utopian banquet for all the world. Well into the eighties, federal employees often held banquets in the Xingu Park for "illustrious" visitors of faraway lands. Seeking the primordial matrixes of their long forgotten pasts—the evolutionary stance they cannot get rid of—they feasted on imported wine, grapes, Swiss cheese, and caviar brought in by the country's air force planes for an adventurous day among "savages."[4] The gastronomic ceremony ended with the Indians avidly cleaning up leftovers off their "guests'" plates while panties were distributed to women, sugar-candy to children, and soccer T-shirts to men.

"You feed us leftovers," says Juliana, "leftover food, leftover land, leftover clothes, leftover medicine."

"You?" I ask. "Why do you include me?"

"Look in the mirror, the mirror."

Tired of working for the SPI, which demanded much from the Indians but did not honor its promises of industrialized gifts, in the mid-fifties Sabino was employed by rubber gatherers themselves as an "inspector." He was told to kill rubber tappers who did not obey him, in exchange for plenty of clothing, tools, a hammock, blankets, and food—powdered milk, rice, beans, tomato paste, pasta, coffee, and sugar. Several years later, in the early sixties, he declined Prepori Kayabi's invitation to move to the Park. Prepori had recently abandoned his original territory in northwestern Mato Grosso, where the Kayabi were constantly being harassed by incoming settlers, for the "safety" of the Xingu boundaries and the "generosity" of the Villas Boas brothers.[5]

I didn't want to move to Xingu since my bosses gave me plenty to eat, and I didn't know what it would be like in the Park. But Prepori kept telling me how prosperous I would be and how great it was to live in such a place. There were many Kayabi living there already, too. He kept trying to convince me for years, and I finally gave in. But when I arrived at the Diauarum Indian Post, I did not see anything, only four small houses, a boat, and an airstrip. No food, nothing.

> Hu ê hê, hu ê hu ê . . .
> [Speech]
> Hu ê hê, hu ê hu ê . . .
> [Speech]
> Hi ê wá, hi ê wá, hi Hi ê hê ki ê, hê ki ê, hê ki ê . . .
> Hi ê hê ki ê, hê ki ê, hê ki ê . . .

Sabino blows the *yawacan* (jaguar bone whistle) and sings, in trance. "He is calling *mamaé* to cure you," says Juliana.

"But I am not sick," I reply. "Am I?" "You have no God, you told us. Where is your *aéan*, your soul?" asks Juliana. "It is probably wandering up there," she replies, pointing to the sky. To my understanding, she was pointing to the domains of death and the beyond, to the unreachable and the intractable I had never dealt with consistently. "Do you ever pray?" she asks.

"I . . . I . . . I . . . " My words are caught in this other scene, one that frightens me to the limits of the impossible. "My legs," I mumble, "my legs are paralyzed. I cannot move."

"Your *aéan* has been captured, too, Mariana; *Añang* has gotten to you as well."

Matareiyup, Sabino's son, translates the chant:

> I see everywhere
> There, the path
> Where I heard a voice
> A low voice
> In the middle of the path there is a hawk
> Where I almost got lost
> First I didn't know the way
> I am glad I had food
> I almost forgot how to pray
> I almost died

I am glad nothing happened
I almost left you

Thus I remain
Thus I tame the ferocious animals
Thus I tame the hawk
Thus I tame all the ferocious animals for us
When the animals go wild
I tame them with my prayer

Everything has been tamed
It went all right
I will pray for us
Everything will be all right between us.

"He is taming the spirits," says Juliana. "He is naming them one by one: *Oua-capeun* [Black Wood], *Yurupininun* [Mouth Painted with Black Dots], *Aucoun* [Black Hair], *Uyupchinin* [Noisy Arrow], *Caawot* [Dark Woods]. . . . Hold on to the mirror; don't let go of yourself."

First I will tame them for you
I will pray
Hold on tight
I will get it for you
I will get the spirit
For you
You will be careful
I myself will get it
For us
Don't let him go away
I will tame
I will grab
It's hard to believe
I am talking with the spirit.

"He is trying to extract the *mamaévévé,* the magical object that's been caus-ing you harm, from your body."
"My legs," I murmur. "I cannot move my legs."
"Just keep still."

Everything will be all right
Will be prayed for
That's what I was told

"I will cure"
That's what he told me
"Everything will be all right"
That's what he told me

"Don't be afraid
We will cure
I am here
I am taking care of
What is sick"
That's what he told me
The one who will cure with me.

Hi ê wá wá, hi ê hi ê . . . Hi ê wá, hi ê wá . . . Hi ê wá wá, hi ê wá . . .
Hi ê, hê ki ê, hê ki ê hê hê . . . Hê ki ê, hê ki ê, hê ki ê . . .
Wá wá wá wá .wá wá wá wá wá wá wá wá wá wá . . .

Juliana lights a cigar and blows, blows my legs, my hips, my arms, my life-
less limbs. I am no longer in control; my dissociated self . . . father . . . husband . . . children . . . fright . . . corpses . . . body . . . hypnosis . . . spirits
. . . blood . . . ambitions . . . memories, reminiscences. The mirror, where is
the mirror?

"Here, this is what was troubling you," says Sabino, handing me the cracked
mirror I had been desperately clinging to.

"If it was troubling me, why do you give it back to me?"

"You've stared through it long enough. Now I can go on."

From 1966 to 1974, Sabino worked for the Villas Boas brothers in the Xingu
Indian Park, which had been created by the Brazilian government in 1961. He
served them as a cook, housekeeper, and gardener and worked as a nurse and
a "pacifier" in different "wild Indian attraction fronts." In 1968, for example,
he was sent, against his will, to "pacify" the Indians who lived in the south of
the Park, who were known to be "sorcerers."

I myself was scared to death. Nobody liked being there. The Indians could bewitch my children, and my wife didn't want to go. But Cláudio Villas Boas said the boat was already waiting for me. He kept insisting that I would have plenty of milk there to feed my recently born twins. He was so mad at me that we decided to leave. But it was the same; as we arrived we saw nothing. The post was a wreck, and I was supposed to fix it up and keep the Indians that lived there quiet.

unveiling traumas in mirrored images

The exquisite sensations I experience hardly allow me to take notes. My whole body shivers, my head spins, and I can barely keep up with the shaman's speech. There are too many questions I want to ask. "Why me?" I ask. "Why the mirror?"

"I have cured you. You are responsible now, you. You have tamed the mirror. Tell your children the history, your own history. Show them the mirror. Do you have mirrors in your house in São Paulo? You should."

Only in his late forties, when the Villas Boas retired and left the Xingu Park for good, Sabino was able to be with his own family and kin, plant his own gardens, fish and hunt for his wife and kids, gather around the fire at night to sing and tell stories. A political leader of a faction of the Kayabi population, he was always known for his generosity toward other Xingu peoples as well. When the Panará Indians were brought into the Xingu Park in 1975, for example, and were too weak to work,[6] they found food and shelter at Sabino's village. But only at the age of sixty-two, after having his soul abducted, was Sabino finally able to *rest*.

When I look at my life from above, watching myself go through such hardships and tragedies, I realize why I was constantly mad at the white men, at myself, and at my own people. So much anger in our lives, so much suffering. And now that I'm able to look back from above, now that my legs can't take me anywhere and that my arms can't do any work, I feel free. Free to be a real Kayabi, to dream of those bad days that are gone, and to think of my son Yawariup as the chief of the Diauarum Indian Post, elected by the Indians. The first elections in the Xingu Park!

Do *you* know why I insist? I insist because in my flights I cannot dissociate mirrors from my own representations of myself or ourselves; I cannot help associating mirrors with the uniqueness of humanity. Like an evil eye, mirrors unleash the imaginary, thrusting meanings between the movements assumed in the image and the reflected environment. I insist because I have identified myself with ourselves, precipitating the *I* within the *me,* through a mirror-stage. Not only myself, *our*selves. Sabino and Mariana. Both *I*'s as ideations, idolatries, ids, idiosyncrasies—*I*'s disguised as *we*'s in the process of becoming.

Notes

1. The Tuiararé Village, named after the Kayabi Creator, was built in 1987 as a main site for the reunification of several families that lived along the Xingu River in the Xingu Indian Park. In 1990, there were 526 Kayabi living in the Xingu Park.

2. The Kayabi were brought into the Xingu Indian Park in the state of Mato Gross, by the National Indian Foundation (FUNAI) in the fifties and sixties from their original territories in the northernmost area of the state of Mato Grosso and from the southernmost part of the state of Pará.

3. The abduction of the soul is the principal means by which a Kayabi becomes a shaman. This disembodiment, albeit involuntary and eventually fatal, precipitates communication between human and spiritual beings. The Kayabi cosmos, as for other Tupi indigenous groups, is organized in a series of "layers," or realms, each of which is inhabited by different creatures: human beings and different spiritual beings both malignant and benign. Although human beings and spiritual beings possess similar attributes— they can speak, hear, and sing, and they have families, material goods, and pets—the former possess a soul (*aéan*) that inhabits a body (*aipit*), whereas the latter may, in their invisibility and "spirituality," take on any form or come to inhabit, in trying to become visible, any object, animal, or human.

4. I witnessed such banquets between 1980 and 1984 in the Xingu Indian Park, when Brazil was under military dictatorship. FUNAI was thus controlled by the military, and it was they who engaged in these gastronomic feasts.

5. Cláudio and Orlando Villas Boas worked for almost thirty years in Central Brazil as FUNAI employees, "pacifying" Indian groups. Although they have achieved worldwide fame, the Indians of the Xingu Park have recently stepped forward to reveal the colonizers' close association with the corrupt military, land usurpers, and other anti-Indian interests.

6. Of the nearly one thousand Panará contacted by FUNAI in 1974, approximately seventy were still living in 1975.

If the Raven Should Croak before I Wake

Gary Nabhan

I saw it from far away, from a promontory, like a dream: a raven falling from the sky. It fell to the barren ground in a circular clearing which looked as though it had been plucked of all its grass cover by a swarm of ants or, perhaps, by humans. "Wait!" I cried, but it lay still, and I grew sad that its life and mine might not ever converge.

Suddenly, I found myself running down from the top of the mesa toward where it lay, down the outwash slopes of gravel left over from the Pleistocene, down zigzagging trails shaped by the running of mule deer, down their switchbacks with my mind's eye all the time fixed upon the clearing with a black shining heap in its center. I hit the bottom of the slope, and without breaking speed, meandered through the tall clumps of sacaton grass, scaring up coveys of quail and even a few slumbering peccaries as I went. Finally, I came into the clearing and chugged to a stop, streams of sweat blurring my vision, burning my eyes, as I knelt over the limp, warm body shrouded in black plumage.

No blood, no signs of parasites nor of injury inflicted upon it by some other species, no visible causes of death. The raven had expired in flight. I inhaled. I had come upon it so freshly that it did not yet carry the smell of death along with it. I cupped its torso in my hands, and one last bit of breath passed from its breast.

The next moment I recall being with the raven was that night, at a campsite where a number of us huddled around a large black pot on a bonfire. I was adding the raven breast meat to the pot, carefully holding the dead bird's neck and back in one hand and scraping its flesh with the knife in my other hand. I could not tell how dark the meat was for sure, but as it fell out of the campfire light into the shadows of the pot, it felt as if we were cooking up the very night itself. The meat sizzled and sputtered in a thick, chocolaty stew with juices, onions, and herbs until it looked as though it were substantial enough to feed us all.

We moved the pot away from the reddest coals to let it simmer awhile longer on the edge of the fire. We called everyone together, chanted a brief blessing, then dipped ladle after ladle into the raven stew. The entire community commenced to eat.

While looking around through the smoke of the bonfire, I noticed that my neighbors were reacting to the stew with something other than pleasure or distaste. At first, I could not figure it out: Some of their mouths puckered closed, then reopened soundlessly, like the mouths of fish seen swimming and feeding underwater. Others seemed to bubble forth with a mad chatter of cacophonous sounds, which became more and more melodious as well as more profound. Those who had previously been soft-spoken—or perhaps reluctant to make any public utterances at all—were suddenly speaking in the most beautiful and articulate of languages. And those who had been loquacious for as long as I had known them—the ones who had always dominated conversations with their polished rhetoric and wit—were suddenly struck dumb. I cannot fully describe how it felt to be in the midst of such a transformation, except to say that I heard it with my own ears and saw it with my own eyes: The meeker voices did indeed inherit the earth. They did so while the others ate crow and listened until their ears and tongues burned.

Outside the Cage Is the Cage:
An Interview with Errol Morris

David Rothenberg

It's not so easy to find Errol Morris's office. For one, it's called "Fourth Floor Productions," but it's not on the fourth floor. Then you enter, and the first room looks like some sort of delivery service, a long Formica counter with packing materials. The back rooms contain computers, editing machines, film clips, boxes of CDs. But the decor is antique, resembling some seedy detective office. In the boss's room are piles of unusual books, all left over from various unfinished projects: Six books on the life of the philosopher George Berkeley, a coffee-table-size volume entitled *Alien Discussions,* the transcripts of conversations between abductees and their captors. The phone is constantly ringing. There's a big Elsa Dorfman polaroid portrait of Morris's eight-year-old son, "dressed up as a genetically engineered food source." Morris is hunched over a bit in his office chair at the computer, looking like a beleaguered college administrator, as he talks emphatically on the phone. "Auschwitz? I've never been to Auschwitz, but I've always wanted to go. Can you get me there?"

This is the nerve center of one of our most talented and original filmmakers, best known for his 1988 feature documentary *The Thin Blue Line,* probably the only movie to overturn a murder conviction. Then he made a film of that heavily purchased and little-read Stephen Hawking bestseller, *A Brief History of Time,* shown to wide acclaim in 1992.

His latest film is *Fast, Cheap & Out of Control,* and it is all about humanity's relationship to the natural world. It took Morris many years to get funding for this movie because it is almost impossible to summarize. It's the story of four very disconnected characters, each with his own unusual obsession with human nature and the rest of the world—just our *Terra Nova* bag. First is Dave Hoover, a lion tamer who dreamed of being just like Clyde Beatty, the star of hokey thirties adventure movies like *King of the Jungle*. Next is Rodney Brooks, a robot scientist at MIT who makes artificial creatures that look and move as if they were alive. Then comes George Mendonça, sole caretaker and gardener of Green Animals, a topiary garden in Newport, Rhode Island. Finally, there is Ray Mendez, an enthusiast for the African naked mole rat, the only mammal that lives in a manner reminiscent of social insects like termites or ants.

The story of their four disparate obsessions with a different aspect of nature is interspersed with footage from classic Clyde Beatty movies and scenes from a modern-day circus. The cinematography is by Robert Richardson, who shot the visually fascinating but morally bankrupt film *Natural Born Killers*. *Fast, Cheap* is a lot more fun; it's deeply provocative around that shaky balance between nature and culture, offering more questions than it does answers.

I had heard that Morris, like a small coterie of other innovative filmmakers, had started his career in philosophy. I had to find out the truth.

Terra Nova: Rumor has it that you dropped out of four philosophy graduate programs. Why?

Errol Morris: For a while I was convinced that philosophy was a place where I could remain interested in ideas and write about ideas. But things went from bad to worse. I was first at Princeton. There was a lot of exciting stuff going on there. Hilary Putnam and Saul Kripke were writing on realism, naming, and reference, while Thomas Kuhn was presenting his theory of changing paradigms in science. I was very much interested in the history and philosophy of science, and Kuhn was my advisor. Unfortunately, I could never get along with him. The philosophy department made him nervous and he wanted little to do with it. At one point he even told me that he would throw me out of Princeton if I continued to take philosophy courses. And it came to pass.

Terra Nova: What was his problem with philosophy?

Morris: I like to think he somehow knew that his book *The Structure of Scientific Revolutions* made no sense. In each new edition he would move closer and closer to a position like Putnam's and Kripke's and away from what he had been saying when the book was first written. Honestly, it was never altogether clear to me what he was saying. The book struck me as deeply confused. This whole issue about incommensurability of meaning was puzzling because it seemed to me that if true, how would history of science be possible. If we are all prisoners of a certain way of seeing the world, how is it that we can liberate ourselves from that prison and see from one world of meaning into another?

Terra Nova: You're asking me?

Morris: If the argument is just a sociological one, namely that social factors change the way we see the world, I think that is unexceptional, and not terribly controversial. If there is this radical incommensurability between one way of seeing the world and another, between one "paradigm" and another, that seems to me something else altogether. How then is science or real knowledge possible? Well, that's about as much as I want to talk about this sort of thing. I'm out of that racket now. I mean, someone just asked me today what my film has to do with the Heisenberg uncertainty principle. I told him, nothing whatsoever.

After Princeton, I was at Oxford briefly. Then I was at Harvard for a year in 1970. Putnam was giving a course on Gödel, which was more like a course on Mao. This was a time of deep social unrest in American universities. Then I went to Berkeley. I found Berkeley to be the least congenial of all the places I went to. The department was so truly bad. There was a feeling that since no one was really a first-rate philosopher, they would just sort of quibble and pretend. It was moribund, and unpleasant. That's when I started going to movies.

Terra Nova: Do you feel that your films reflect a philosophical point of view?

Morris: Sure.

Inexplicable Insanities

Morris: Before my career as a professional philosopher-wannabe came to an end, I wanted to write a thesis on the insanity plea. A book by Herbert Fingarette had come out, on the philosophy of the insanity plea, and it was something that had interested me very much.

I had this interesting theory, which you can probably see reflected in my movies as well. When we talk about insanity, we talk about it in two different senses. We talk about insane people and insane acts. The word is used in two radically different ways. There is an insane state of mind—for example, we say: "That person's insane." And then there are insane actions: "Shooting the President was insane." Something is done, something happens, but we can't understand why anyone would do such a thing. My theory: In an effective insanity plea, you should make the act seem as strange and inscrutable as possible. You shouldn't offer a psychological or psychiatric explanation. You shouldn't try to explain why things have happened.

I started going to the trials of a number of mass murderers in the Santa Cruz area. In the early seventies there were three very famous murderers in the Santa Cruz vicinity, and I talked to all three of them. I went to part of one of the trials, the trial of Ed Kemper—I think the most fascinating of the three. Ed Kemper had been put away for killing his grandparents when he was twelve, thirteen years old. He was released several years later, and he went back to live with his mother, who worked in administration at the University of California, Santa Cruz. He began to borrow her car, with UC stickers on it, and he would pick up hitchiking coeds. They would get into the car, he would drive them to remote locations, and he would kill them.

Terra Nova: So he got out of juvenile detention and killed again?

Morris: Yes, and that's one of the reasons why this case received so much attention. All of those professionals at Atascadero, where he had been incarcerated, decided that he had to be released.

Terra Nova: Where is he today?

Morris: He's, I believe, still at the Vacaville medical facility in California.

Terra Nova: Deemed to be insane, or just incarcerated?

Morris: Well, I would say that if that term has any meaning at all, then Ed Kemper is truly, deeply insane. During the trial there was a psychiatrist, Joel Ford, who testified that Ed Kemper was sane, and not only was he sane, he wasn't even neurotic. What came out was that he was suffering from an affective disorder that made it difficult for him to relate to other people.

Terra Nova: That's all?

Morris: The example was given that when he was killing these women he would talk to them and say things like, "I hope this is not too unpleasant for you, I hope you are not too uncomfortable," and so on and so forth. Now, the doctor took this to be proof that he was completely aware of what he was doing but was unable to emotionally connect with his victims. Hence sociopathic, psychopathic, but not psychotic.

What really fascinated me was the epistemic part of it. Where did the doctor get all of this information? He got the information from Ed Kemper; it was the only place to get it from since the victims were dead.

Terra Nova: So you're saying that Kemper could have just been telling a story?

Morris: Maybe his fantasy is, instead of being completely out of control, he was completely in control, so completely in control that he could talk to his victims dispassionately while killing. The psychosis of being completely in control when you're completely out of control. That formed a lot of what has fascinated me through the years.

Terra Nova: The gray line between in control and seeking the out of control—how we know what's really going on?

Morris: How do we know what ultimately motivates people? Whether people, when giving descriptions of the world around them, are really providing descriptions of the world around them or some dreamscape that they have constructed for themselves? Or all of the above.

Terra Nova: That's definitely part of *Fast, Cheap & Out of Control*. You've got these people with their obsessions and their ways of looking at the world, and you try to show how they connect to one another without forcing the link.

Morris: This is part of every single film I've made. Do we exist in some kind of strange cocoon? What is our connection to reality, to the world? I like to cite the first lines in *Vernon, Florida,* when Albert Bitterling says, "You mean this is the real world?"

Whose World Is the Real World?

Terra Nova: The real world has to be somewhere else because in *Fast, Cheap & Out of Control* you get a real sense of how each of these people describes the events of the world around them, but you don't necessarily know so much about what these four characters' lives are like. You mainly learn of their obsessions.

Morris: Well, my interest in the film is not in the objective details of what they do, but how they describe it, how they see it. Recently there was a series

George Mendonça: *This is all done from memory. You know what an animal looks like, and so you just start making an animal. They constantly need repairing all the time. I like the bear, myself, because it gives me less work. . . . But it took me fifteen years to build that bear. I won't live long enough to make another bear like that one.*

on public television about Stephen Hawking's universe. I watched part of it and it struck me that this was exactly the kind of thing I didn't want my movie to be. This was science pedagogy at its worst. You see, there is this confusion: People assume Stephen Hawking's book to be a How-To book—how to learn advanced physics in three simple lessons. A sort of Quantum Gravity for Dolts. But that's not what it is at all. I've always looked at this book as a romance novel. Because the book is chiefly about Hawking's personal investment in his work, about how he sees the world.

Terra Nova: After Stephen Hawking, the characters in *Fast, Cheap & Out of Control* seem less extreme by comparison.

Morris: These four guys are eccentrics, but I would say, yes, they're less extreme. They seem like you and me. There is a strange populist idea at the heart of the film.

Terra Nova: There is something very realistic about it, and for our readers, the real questions are what are you trying to say about the relationship between people and nature, animals and the environment, and our stories of what nature is about. And each of these characters brings a whole new spin on what this is. I guess that I would like you to say something about what you are trying to say about the human world and the natural world.

Morris: When you hear these four people describe their relationship to nature, you realize that there is no place for any of these characters in this world. For Hoover the lion tamer and Mendonça the topiary gardener, their worlds are winding down and have nearly run their course. Maybe there was a time when it seemed like the garden could go on forever, but no longer. And the circus will never seem as romantic as it once did. Here you have these two characters talking about worlds that may not even survive them.

Brooks and Mendez, on the other hand, are obsessed with worlds that could very well survive without them, worlds, in fact, that could last in some form long into the far distant future. The mammals who live like social insects or the silicon-based robot world might well live on beyond humanity. But these are worlds that inherently exclude these two guys—they exclude all of us.

Terra Nova: Where's the right place for humanity in these obsessions?

Morris: Where's any place for us? There's a sense of dislocation that I like in the movie.

Terra Nova: The film is easy to watch, though. That too may be disarming. It's fun, and among the world of collage movies this one's so much more fun than many. I mean, the world of the circus and the world of the 1930s adventure movies you use in the film, it's sort of welcoming; you want to be there.

Morris: You mean *King of the Jungle* and *Zombies of the Stratosphere?*

Terra Nova: What happened to the little Nature Boy from *King of the Jungle?*

Morris: I don't know, although according to Dave Hoover, he might be around someplace. He'd be in his seventies now.

Terra Nova: You didn't try to find him?

Morris: I'd like to try to find him. Nature Boy, please stand up!

Terra Nova: Do you think that your four characters share a sense of irony about what they believe in, or are they sort of naive in their convictions?

Morris: I think that no one is entirely naive. There are all kinds of levels of irony in *Fast, Cheap*. Dave Hoover's first lines in the film are incredible. He's not telling us he wanted to be a lion tamer as a little boy. He says, "I wanted to be a Clyde Beatty"—Beatty, the hero of these wild jungle films of the thirties. Hoover didn't just want to be in a cage with wild animals, he wanted to enter into this virtual landscape, into this artificial jungle on celluloid of the world of Joba, with the Batman and Nature Boy and the princess and the whole deal.

Terra Nova: It never meant that he wanted to go to the real jungle and have adventures, but only the fantasy of adventures?

Morris: It's several layers removed. Clyde Beatty, in turn, was already recycling some nineteenth-century dream. You know, with the pith helmets and explorers, like at the source of the Nile or people facing wild beasts in the jungle with impunity.

   In an early version of the film, I had robot scientist Rodney Brooks talking about seeing *2001* as a child and deciding the computer Hal was his favorite character.

Terra Nova: I think I identified with Hal too.

Morris: Well, Hal was in a way the one truly likable character. Maybe his tragic flaw, like all of us, was that he was just unable to accept death.

Terra Nova: Yes, I think so. And yet he still remains a more human character than any machines really are. Rodney Brooks doesn't think his robots will replace humanity, as Marvin Minsky seems to intimate. It seems that Brooks's greatest contribution is to point out that his machines don't have to know what they are doing. We just prepare them to respond and look at their walking, and there isn't a collective intelligence behind it. It just responds, and maybe we're more automatic than we think; maybe we don't need the self to walk around. He does seem to believe that *nature* is quite mechanical, though.

Morris: Maybe all of these ideas of soul, self, and individuality are higher-order concepts that we invent in order to explain those hidden, simple feedback-loop mechanisms, hidden away.

Terra Nova: I thought Mendez was most enigmatic as a character in that I couldn't figure out whether he's a scientist or a photographer or a mole rat enthusiast. He didn't convey an expert mentality.

Morris: His real job is that he's an insect wrangler for the movies, for commercials, and for television. Like, when you see a lot of roaches in a movie, chances are that Ray Mendez has been involved in wrangling those roaches. He did the moths in *Silence of the Lambs;* he did the roaches in *Joe's Apartment.*

Terra Nova: It's interesting that you decided not to tell us that in the film. It makes it more mysterious. I thought that he was either a scientist or a photographer.

Morris: He also does a lot of photography, a lot of mole rat photography. He also builds exhibits for various zoos, and he's building a whole mole rat exhibit for Disney.

I like very much when he talks about the mole rat exhibit as a doll house, like this idea of control. You know, if you wanted to see such and such, you can do this and that, and so on and so forth.

The film has a whole set of arguments. It is about people creating worlds, sort of like worlds of animals, worlds involving some kind of life. It's sort of interesting that Dave Hoover has one of my favorite lines in the movie, "outside the cage is the cage; inside is their world." Of course he's not just talking about lions and tigers; he's talking about all of us. And Rodney echoes that sentiment quite well when he says that he may know in his heart of hearts that all it is in the end is simple feedback-loop activity, but you can't limit

Dave Hoover: *This is where they sleep. . . . This is where they drink. . . . All their creature comforts are connected with that cage. They don't particularly want to leave it. They don't consider themselves caged. . . .* Outside the cage is the cage. . . . *Inside is their world.*

your life that way. It becomes necessary to sort of occupy yourself with what may in fact be some fictive world, but a world that is inhabitable.

Terra Nova: You have to live beyond the theories even if you believe the theories.

Morris: Or else you have to live somewhere other than reality in order to survive. I think that's it in a nutshell.

Terra Nova: But is that the particular obsession of the four characters, or is it the real life that they do not speak about? Which is the nonreal; which is the real?

Morris: Well, I don't know.

Terra Nova: Exactly.

The Thing About Animals

Morris: Well, I do think that this thing about animals, about nature, runs through the whole movie. I mean, it becomes an obsession of the movie itself. But I can't really say why. I mean, there's part Doctor Frankenstein in it, that desire to create and control life. And part of it is this idea that Rodney expresses, that by manufacturing life, or manufacturing some facsimile of life, we can come to understand it or understand ourselves—this idea of studying life as a mirror, some strange, metaphorical mirror of our existence. All of that runs here and there in the movie.

Terra Nova: Yes, there seem to be these points where all of a sudden the characters get excited about how insect societies are fine, but when you have mammals that live like insects, it's one step closer to us—like the mole rats. And then you have these hedges, which are hedges, but then if you take the plants and make them like animals, you know, why not make them like fire trucks or skyscrapers? It's something about turning plants into animals. I mean, are all topiaries animals?

Morris: No, in most topiary gardens the hedges are in some geometric form— spheres and cubes, and rectangular solids, and spirals and cones. But the idea of animals, this kind of strange garden of Eden. The movie moves from some version of the garden of Eden through to future worlds.

Rodney, in part of the interview that did not actually make it into the film, talks about the dangers of simulation; it's one of his big shticks. He says that you actually should build these things rather than create virtual creatures that really just existed in some kind of program.

Terra Nova: Does he ever say anything about how our popular image of the robot is so much more humanoid and complex than real robots? Real robots are all around the industry now, doing all kinds of impractical tasks for people, but they still don't look like *Lost in Space* or R2D2. Is he trying to maintain the romance of the humanoid robot—malicious robots like that vicious creature that gets electrocuted against the pipes in the film?

Morris: Well, he has a humanoid robot now that he's constructed, called Cog.

Terra Nova: So he's moving in that direction.

Morris: Yes, he's trying to raise it as a child.

Terra Nova: How is it working out?

Morris: We haven't had a chance really to talk.

Terra Nova: Does he think of how the movie relates to his work? I think the movie conveys this sense of human intelligence beyond the irony. There's a way of human understanding that by putting all these different things together and assessing these images from the past and the present and the future, then it becomes a human activity, and that's what makes us human.

Morris: I think that's true.

Terra Nova: That we can make sense of all this craziness even though people are trying to say that we're either mole rats or machines or things like that, but no, we can make films like this, and we can watch them and get a grasp of this other kind of intelligence.

So, with these characters and their relations to the natural world, is there any vision that you are trying to portray of how we should relate to nature?

Morris: Well, I think there is a perverse paradox in all these stories. It's like Ray Mendez. His obsession is with the other, and he defines the other as that which has no relationship to who or what we are. There are animals, such as termites or wasps, that exist completely independent of us.

The Competition

Terra Nova: Have you seen any other of the weird nature movies that have come out? Like *The Congress of Penguins?*

Morris: No. Is that good?

Terra Nova: Oh, you should definitely see this one. It's a German film, and it begins with a shot of penguins sitting on the ice. Actually, it begins in darkness, and there is a voice-over. The guy says, "I was in a noisy, European city, asleep in this hotel. I dreamt I was out in the ice, hundreds of people were standing in front of me, staring." Then the scene switches to Antarctica; this guy's sitting in a down jacket, hundreds of penguins looking at him. "They had something to tell me." And it sort of goes on in this strange vein. It's actually a Holocaust story in which the penguins are somehow wanting him to atone for the fact that penguins were tossed into stoves as fuel by Norwegian whalers.

Morris: Really?

Terra Nova: At the turn of the century at a whaling station called Grøtviken, the whalers tossed the penguins into the flames as fuel as if they were nothing, and that's how they fueled the fireplace to get warm. There was no wood, just penguins. They tossed them in, burned them, because they are all fat. So there was this horrible vision of what went on that begins this strange film about Antarctica and scientists and birds. A lot of the footage looks like normal nature footage, but the story continues insanely in this direction. You should see it.

Morris: It sounds good.

Terra Nova: Then there's *Microcosmos.*

Morris: I saw that. I call that *bugsploitation.*

Terra Nova: Bugsploitation?

Morris: Yeah, to the best of my knowledge, none of those bugs were compensated for their participation in the film.

Terra Nova: It's a French film—not just insect battles, but bugs making love.

Morris: I didn't think that it was all that weird.

Terra Nova: It wasn't weird enough?

Morris: It didn't seem weird at all. In fact, it was boring. It seemed to banalize the world rather than make it of any interest.

Terra Nova: Although the dung beetle acting out the myth of Sisyphus was pretty good.

Ray Mendez: *You know people are afraid of new, different, and strange things, but to me these crea-tures are not to be feared. They are to be wondered at, looked at, and explored, perhaps communicated with. See, if you can get the moment when the animal is actually looking at you, when you feel there is a moment of contact . . . It's not something that happens every day. You have to go out and look for it.*

Morris: That was okay.

Terra Nova: And then when you find out it was all storyboarded and set up—that's much more like the Ray Mendez bug talent agency or something.

Morris: Yes. There should be a whole genre of oddball nature films, but it seems that the stuff that makes it on American television is subverted by this idea of science instruction, and they have a kind of sameness to them.

Terra Nova: Yeah, that's true. The music is the same. There are convoluted stories about what the animals are supposed to be doing.

Morris: And on the other hand, there is a sort of new genre that's being cre-ated at Fox, in which animals attack. And now there's the new version that has come on the air in just the last couple of weeks, of when animals *swarm*.

Terra Nova: Why do you think that these are getting popular?

Morris: I'm not exactly sure. Maybe we are in some kind of uneasy relationship with the world out there, or there's a level of guilt that's building up to where we feel it's going to be payback time. Just think of it, gangs of marauding penguins immolating unsuspecting Norwegians.

Terra Nova: Well, there's a whole tradition of that, of course. The fear of nature—I think it's a lot older than the desire to save it.

Morris: Yes.

Environmentalism in Film

Terra Nova: There's the *Star Trek* film where they come back from the future to save the whales. Remember that one? They need to go back in time to San Francisco in the mid-1980s because whales were very important, and this was a last chance to save them. So they all come back to save the whales. Do you think your film has an environmentalist component?

Morris: At the New York Film Festival, someone stood up in the audience and asked how I felt about the treatment of rare animals in the movie. And I said, "I assume you're talking about my four protagonists. I tried to treat them as well as possible." Then I told a story that comes from Dave Hoover about how the circus was picketed by animal-rights activists. This woman came up to Dave and said, "Why don't you send that lion back to where it came from?" And Dave looked at her and said, "Now, why would that animal want to go back to Erie, Pennsylvania?" You know, there's a project that I've wanted to make for years. It's the feature film that I may try to make next, if I can ever sit down and think of rewriting the script: the story of a dog that was put on trial for murder.

Terra Nova: I can see the ads already: Two people were at the scene of the crime—one is dead, the other is a dog.

Morris: What is really so deeply fascinating is that the story is a tabloid story, which is far from a joke. We are taught that this is part of the law, that the law applies only to persons because there is an intentional or mental component to a crime, and that therefore the law does not really apply in the case of, say, dogs and cats; it doesn't apply to animals. But here you have a story where, in order to really understand the story effectively, all of these concepts play a role. The whole story turns on the issue of whether or not in fact this bite was intentional. It's not devoid of meaning at all; in fact, it has real meaning. What really did happen? Did the dog wantonly attack his owner's mother? Was it an accident? Was it an unintentional homicide?

Terra Nova: Is the dog still alive?

Morris: The dog is no longer alive.

Terra Nova: Was the dog quickly put to death?

Morris: No, the dog's life was saved, but the dog died about six months afterwards.

Terra Nova: Of grief?

Morris: Don't know. It could have been grief. Perhaps it was the kind of pain that comes from the stigma of a false accusation.

To Keep or to Release

Terra Nova: You have mentioned that Hoover and Mendonça are preservationists and that Mendez and Brooks are either revealing new things in nature or reinventing nature. Is there an environmentalist message here? We have a lot of environmentalist readers, in the larger sense. Not those who will necessarily picket zoos, but those who want to know what about nature is worth saving and paying attention to, and what about it, how much of it is a human construction that changes. This is a big issue among thoughtful environmentalists—where's the nature out there that we are a part of? What should we preserve? Your film certainly deals with these issues.

Morris: I think nothing ultimately will be preserved. I went to the Museum of the Rockies with my son, put my finger in a triceratops brain cavity, and said, "Someone might be doing this with me someday." And the curator said, "You should be so lucky." Only the privileged few of us get to be fossils.

The world is about change, catastrophic change. We don't know if we're part of some design or interfering with some grand design. I remember Stephen Hawking telling me a story years ago about why we don't hear from civilizations in outer space. Civilizations destroy themselves when they reach a certain point. Here we are with unchanged jungle DNA, unchanged for a hundred thousand years, with weapons of mass destruction. The outcome cannot be a good one.

Now he has modified that view and said that the only way we can survive is by changing our DNA. We will start modifying our genetics in order to ensure our own survival and the survival of life on this planet.

Terra Nova: At the same time, do you think that *you* are nostalgic for a certain kind of nature or period of time, just like the characters in your film?

Morris: I think that there is some truth in that. I have spent a lot of time in wilderness areas. I was a rock climber, and I've lived in Yosemite Valley, the Tetons, the Canadian Rockies, as a climber.

Ever since I was a little boy I spent a lot of time hiking. I took my son on the Long Trail in Vermont recently—I live outside North Bennington—but it seems so different nowadays. Being on the Long Trail, we got to a shelter, walked about seven miles, and saw very few people on the trail. I hadn't been backpacking for at least 15 years. The water's bad. Everybody carries filters and chemicals for it. People would come into the lean-tos with cellular phones. I always had the climber's healthy disdain for backpackers. People were hiking the whole Appalachian Trail. They take on trail names. There was one guy named "Blank." My son said, "I think he's a serial killer. Why is his name *Blank?*"

Terra Nova: This sounds like a Don DeLillo backpacking trip.

Morris: I'm out there in the woods, but it's not quite the Thoreau experience that I remember.

Terra Nova: At Walden, Thoreau was a mile from his mother's house! Is this just what's happened? Is there some authentic way of going out there that you miss?

Morris: I miss the way things were. The old days in Yosemite Valley, when rock climbing was considered to be highly undesirable. No one ever thought of profiting from it or exploiting it. They ghettoized the climbers in a place called Camp Four. That whole world is gone. The people that I climbed with became some of the best free climbers in America in the seventies.

Possessing the World

Terra Nova: What about Mendez, Hoover, Mendonça, and Brooks: What does each of them think about the relationship between culture and nature? Is there a simple answer?

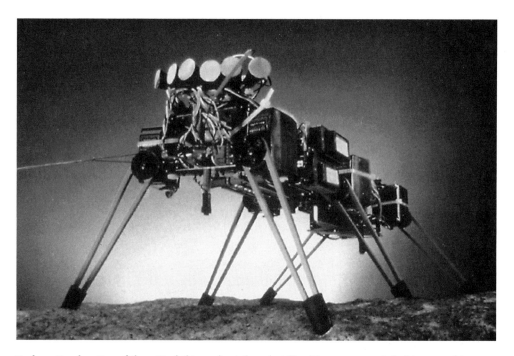

Rodney Brooks: *One of the critical things about the robot Genghis was, you switched it on, and it walked. The walking isn't programmed in. . . . Instead it's all these little feedback loops, and when you put them all together, the robot walks. . . . I switch it on, and it does what is in its nature. . . . Maybe that's all there is. Maybe a lot of what humans are doing could be explained this way.*

Morris: There is a real world out there, something we're trying to *possess.* There's an attempt at a fetishistic connection to the world. Each one of the four wants a special kind of connection with what's out there. For them, nature is a fetish object. Everyone's displaying stuff, like Fuller Brush salesmen—one's got the mole rat circus, another the wild beasts, another the topiary animals, and finally, the robot menagerie. They all say, "Look at them, they do this, they do that!"

Terra Nova: A dog and pony show.

Morris: There's a weird desire in all four of them to possess the world through these things.

Terra Nova: Do you possess the world through your characters?

Morris: I don't know. I think we're all like weird monkeys, involved in weird monkey activities that are only dimly apparent to us.

Terra Nova: Do you want to keep these stories close to you now?

Morris: Well, I made this movie. I like this movie. Part of me is very much like these characters. I'm interested in their worlds. My interviews are a crazy attempt to connect with other people. Interviews are a laboratory of human relationships, the relationship of one person to another.

   People don't always know where to go in my movie—the people in the movies and those in the audience, watching. But I've always felt there's only one real philosophical problem: Where to go next?

The photographs that appear in this article are stills from Errol Morris's film Fast, Cheap, & Out of Control.

Me and Mom and the Bioregion

Jerry Martien

It was a Saturday afternoon in Chicken Beach. After a stormy week, the sun was back and Mom and I were taking it easy. It was not a day that she went to adult day health care, so I had let her laze around and brought her coffee in bed. I'd cajoled and scolded till she got dressed for lunch, and while she cleaned up one of her secret used-tissue nests, I hoovered around enough to suck up the top layers of sand. We got the house cleaned up with hardly any bickering—our housekeeping arguments reminded me of when I was the kid, only it's years later and she's the difficult child now. But today we were expecting company.

It hadn't taken long to realize that my mother needed more social action than her bookish son was going to provide. Part of what seemed to have happened to her—aside from pills, Almaden, grief, and denial—was the isolation of the years during which she and my stepfather had come to the end of their rope. When the old man had a stroke, they'd moved back to her old Anaheim neighborhood, but now it was a dangerous barrio. When they'd taken him to the VA, never to return, she had retreated inward into deafness and dementia—doors locked, drapes shut, not answering the phone or the doorbell for days. Her decline had coincided with my becoming the owner of this derelict beach house, and in the past five years I had made it at least livable. Because she was clearly unable to care for herself, I had brought her north, nearly the entire length of the California coast. It had seemed an inevitable move, and I hadn't given much thought to its long-term effects. But now she was coming out again, recovering some of her old sociable self. She was always cheerful when we were going to have company.

She was sitting downstairs, reading over and over the same page of *The First Third,* with the woodstove cranked up to max even though it was a warm January afternoon. I was upstairs with the windows open, working toward the last chapters of a book about money, which more than anything was probably a tribute to our domestic stability. Between Mom's SSI and a writing class I was teaching at the local college—and an occasional carpentry job here and there— our combined income brought us near the federal poverty level. It was more security than either of us had seen for a while. On the other hand, life with Mom had completely destabilized my closest relationship. It turned out that living with my mother was—no matter what I said—living with another woman. There was some irony in the fact that Joy had been the one to encourage me to buy the house, which had made it possible to bring Mom here. In the ensuing confusion we had broken up twice now, and I had plenty of spare evenings to consider this irony and the additional luxury of talking about things I knew very little about.

When our guests drove up, I met them outside and gave them the brief tour of the grounds. Over the past five years the largest piles of debris had been hauled off or burned, and the scars of foundation work and the neighbors' kids' ATV track had begun to heal—so I could offer them the quick scenic view of our half-acre sand farm. The only real feature of the place was not even on it, but lying over its submerged property line to the west. A good stone's throw across, in a long deep swale between dunes, overgrown with vegetation and half drowned willows, I had hardly noticed it when I bought the place. Deciduous freshwater pond, I said, was its formal name. But you can call it a swamp.

I pointed out to our guests that they were standing on nothing but water and sand—that our squalid little village was strung out on a narrow peninsula, as much liquid as solid earth. It had so far proved resistant to all attempts at discovery or improvement, in an area already so remote that until recently only a few college students formed a buffer between the hippies, Indians, and loggers. Geographically and socially, I explained, we are on the edge of the edge. The next place west is the volcanic black sand beach of Hawai'i. Behind us, a large shallow bay, and around it sloughs and marshes. We are connected to the mainland by a series of bridges across the bay, or by another bridge across a slough that reaches nearly to a river. We are almost not attached to North America. A couple of thousand years ago the river emptied into the bay, probably following a drastic subsidence of the coast. Later, perhaps in another earthquake, the river went its own way north, leaving us stranded here, the next thing to an island. The sign says, "POP 1000 ELEV 2."

Because of its access to water, the peninsula has historically been a place to mill and ship the region's forests, primarily redwood and Douglas fir. For many years, two pulp mills ground up the remaining cellulose and every day dumped forty million gallons of dioxin-laden water into the ocean. Now there is only one mill, bleaching with peroxide, and the scum on the beach is said to be harmless. Anyway, it was surfers, not residents, who complained. These sands have always been a place to dump or bury things no one wanted. It was a convenient site to confine the remaining indigenous coastal people. Later, a place to buy and sell drugs, murder people, party, ride dune buggies, or strip down stolen cars. The law has as little hold as the mainland.

The village of Chicken Beach is the most recent and impoverished of several that sprouted up around the former mills on the peninsula. The coastal forest, a thin strip of mixed spruce and pine, begins and ends abruptly at its north and south margins, cut down for firewood or a better view of its small, sad houses. Our little two-story saltbox is framed and sheathed in odd dimensions of redwood and fir—mill scrap or seconds—and its doors and windows were apparently scavenged from yet another house, fallen down half a century ago.

One morning in the fifties, the local mill workers showed up to find that the owner had emptied the safe, locked the doors, and absconded with all their owed wages. Everything that could be ripped off and carried was soon gone, and the site became a center for vandalism and dumping of every form of trash, a fitting memorial to the years of poverty and depression that followed. Aptly described as Appalachia-by-the-Sea, the land around this community has somehow stubbornly retained its harsh beauty, a stark loveliness that both reflects and sustains our human difficulty. Like the house, it seemed appropriate that Mom and I would end up here.

Standing at the yard's edge, where the blackberries plunged steeply into the swamp, we could look over their winter canes and across the tops of the bare willows to the dune rising sharply on the other side. Myrtle, coyote bush, twinberry, and fern held the steep bank from drifting further this way—or at least slowed it to less than the five-foot-a-year average. Over the dune's top peered the sinister remnant of what must have been an industrial kiln, remnant of a brief and disastrous attempt at economic revival. It looked like a bunker left over from a war on some other planet. Considering the damage both land and community had sustained, it also seemed to belong here. Over it all, carrying three pairs of sixty-thousand-volt lines out to the mills, a steel transmission tower looming like a giant out of *Don Quixote*.

During dry years in summer, I could walk across most of the swamp. But now, a few feet out the dark water was already too deep for chest waders. The footing was a treacherous tangle of live wood and dead and fifty years' accumulation of submerged debris. I'd found the hood of a car in the berry thicket and figured the rest of it might be down in there somewhere. With its rank vegetation and dying trees, branches encrusted with gray-green molds and dripping lichen, the swamp gave off an air of ruin much like our old house next to it. Or the house's inhabitants, for that matter. But in this coastal climate, always somewhere between dismal summer and mild winter, even in January, even half submerged, the willows were red with beginning growth.

In the branches of the willows, stoop-shouldered and red-eyed and irritable, black-crowned night herons let out a cranky squawk at our disturbance and flew to a farther tree. They looked more like they belonged in a pool hall than outdoors—as if instead of foraging in the bay and mudflats every night, they spent their evenings over on Eureka's notorious Two Street. These, I told my guests, are the original chickens for which Chicken Beach was named. Over on the other side of the swamp, out of sight under the willows, a bunch of mallards began quacking loudly, as if one of them had told a particularly good story.

Inside the house, Mom regaled our guests with her usual confabulations. As soon as I introduced her to Trina and Sean, almost before they could start peeling off jackets in the saunalike living room, she started in about how she'd come here from California—for the air, she explained. How she had always loved going to the beach, although she hadn't gotten out so much lately. She spoke as if she were no older than her listeners, who nevertheless sat politely attending to her great-grandmotherly wisdom. I should have told them that when she first went to day care, she would come home expressing great compassion for "those old people." But I'd given up trying to straighten out my mother's stories. Maybe I was beginning to believe them. Maybe this *wasn't* California.

She explained how our family had lived here for generations but then had to move away. The house was abandoned (it had "fallen back," she would always say), but we had returned and restored it. This process of story making is common in the early stages of senile dementia, apparently a natural response to looking into the cupboard of memory and finding nothing there. At first, when she came here to live two years ago, she would simply repeat the same phrases or nervous questions again and again, as if rerunning the same tape in response to a fear of blankness. Gradually, as she adapted to her new environment—a better diet and less wine also seemed to help—she relaxed and became more "at home," and her fables became mostly origin stories like this one. Another concerned the swivel rocker, where she sat by the front window and could see anyone coming. My friend Jack had brought it over when he moved to a smaller apartment, and she'd quickly adopted it. She liked Jack, who called her Mom. But unable to remember these facts, she explained to our visitors that the chair had been left here by a man who was going to move in but then never did. There was an eerie mythic truth in her fabrications, and she'd always been a convincing liar. As she explained to our guests how she used to climb the dune ("the hill"), they nodded with conviction.

Once, I did get her to walk the path through the swamp and helped her up the sliding sand to the top of the dune. That one time, like Balboa, she had seen the Pacific. But I spared our guests this history. My version of the truth had begun to sound to me as demented as hers. I went to find the ragged tennis shoes I wore to the beach.

I had offered our guests the home and garden tour as an example of things I couldn't very well explain. I wanted to show them the land we had been fighting to save from recreational vehicles, a hundred acres of beach and dunes that had been purchased by the state's Coastal Conservancy and put into the care of the local community. And I wanted to introduce them to Mom. It was all part of the explanation. They had driven all the way up here from Davis, through the rain and enormous winds of last week's storm, so Trina could interview Jim Dodge and me for a graduate school research paper she was writing—about our "bioregionalism." What that meant, as far as I could tell (or say), was a very large basket into which we threw all the tangled relations of our lives and locale. Over a two-hour buffet lunch in town on the previous day, we had developed this already murky concept from a patch of valley fog into an obscurity that could easily have swallowed the entire northwest—every person, animal, tree, and rock—and still have room to throw in the spiritual views of Okie Taoism, a pantheist heresy I attributed to Jim's brother, Bob.

It's great to say *bioregionalism* means "living your life in place"—and writing about the experience, naturally—but as a way of translating watershed ecology into daily living, it defies most of our usual categories and often sounds like ordinary horseshit. Maybe that's the present value of the word, I suggested. I had already been to a couple of northern California "bioregional gatherings." The first had been a memorable get-together of friends, old farts who mostly stayed by their own fire and told stories about the sixties. The second gathering was personally distinguished by a transregional argument with my now-estranged companion, as Joy and I began our second breakup since mother had moved in. The quarrel lasted all the way to Mount Shasta, across the entire state and through a September day and on into the mad choreography of putting up a tent in a blundering two-party rage. It would have made a good Buster Keaton movie if we'd had the grace to laugh at ourselves. Although we'd talked in the intervening months, we hadn't figured out how to put the pieces back together.

Like any activity involving more than one person, it seemed that the bioregional movement might prove to be just one more place to bring our cultural and personal problems. Along the way, maybe we would come up with a few innovative responses to some of the worst failings—and hey, we might be part of the much heralded New Paradigm. But it would also inevitably be a haven for the usual reality-starved yuppies and New Agers shopping for an identity. With growing popular acceptance, its name would be reabsorbed by the natural science industry and by planners and bureaucrats who say things like, "a pro-active bioregional approach to mitigation." As the horseshit got buried under the bullshit, maybe we would have to find another word for what we're doing.

You have to admire the nerve of the idea. Just listen to the definition, as it's propounded by the Bioregional Association of the Northern Americas: "Bioregion: A geographical area of interconnected natural systems and their characteristic watersheds, landforms, species, and inhabitory (place-specific and sustainable) cultures." And that's without the adjectival suffix and the "ism." But truly, what other terms are there? If deep ecology is environmentalism with a life, then bioregionalism gives that life a map and says, "Okay, go do it." On what other terms are human beings going to continue *living* on earth? Yet the devil is in that parenthetical detail—"place-specific and sustainable"—expressing whole worlds of things that go wrong, without even considering ordinary human silliness. To say that *this* life is the one to put that into practice is to claim more than anyone knows. To say it of *my* life with Mom seemed a very shaky thing indeed. Sometimes it appeared that "bioregionalism" might just mean doing everything the hard way. And knowing the most you could hope to get—in this lifetime—out of whatever mad plan or project or promise you'd gotten yourself into, was maybe a good story.

We crossed the path through the swamp and climbed the still active slip-face of the dune. From its wide top we could look back across at my upstairs windows. Maybe Mom could see us, if she remembered we'd left. To the west, gray-green and endless, the great curve of the Pacific. We were about a quarter mile inland. I was trying to describe how much sand had moved across this dune in the other night's big wind. That was *wet* sand, mind you. Blowing *north*. And over in the hollow, I pointed to where another remnant spruce snag had broken off. Around us, the debris of the old smelter, or whatever it had been. Something to do with extracting silicone, or some said processing copper ore from upriver. Others said no, it was gold. No one seemed to know. Over there was a green pond, and there a blue pond, they said. And all of it then cut up and run over by four-wheel drives and bikes and buggies.

I had to admit that I was obsessed by damage. I would often come back from the beach more angry and stressed than when I had gone out. I raged. I raved. I demanded revenge for the death of the fine poet Frank O'Hara, struck down in 1966 by a dune buggy on Fire Island. We're talking thoughts of piano wire. But over the years I'd seen a number of organizer/activist/mother's-helper types (and I'd been one of them) burn out around such feelings, and I had to recognize that anger and despair were adding my own soul to the damage. When Mom moved in, and Joy began coming over less, I found myself increasingly isolated with these feelings—and now added to them was an emotional life right out of the Oedipus cycle. Some of my responses were almost that tragic, such as having a lot of coffee with a very attractive family counselor. But some desperate measures actually worked, like going with my friend Jack to a support group for stressed-out care providers. And sometimes, after particularly long walks on the beach, I felt assured that the caretaker was going to get taken care of too.

So I could laugh at myself today, after ranting over fresh signs of destruction, when I had to explain to my ecotourists that no, the damage they were seeing was not caused by vehicles. It's part of our restoration project. We're removing the bush lupine and other invasive species, I explained. And then when I looked more closely—how embarrassing—I had to admit that no, actually all this vegetation must have been torn up by our local teenagers. After each warm rain they suddenly appear, popping up and down in the lupine scrub like large herbivores wearing backward baseball caps. They seem to be seeking a particular mushroom, concerning which they are secretive and sometimes downright furtive. Their astonishing and persistent labor—remember, we're talking about teenagers—may accomplish as much as the CCC and the convict labor we employ in the restoration project. Or they may tear up the vegetation so much that no more mushrooms come, and instead exotic grasses and more lupine sprout in the thoroughly tilled ground, now enriched by the parents of this sturdy legume. It's not as simple as I first thought, the question of recovery, of who and what is foreign or natural, and what life we are restoring to what place.

From the foredunes we looked down on the strand and out at the surf, still wild and crashing from the week's storms. Around us, I pointed out the tough and delicate plants that hold this sloping seawall against all that breaking violence. Sea rocket, beach pea, strawberry, primrose. I showed them the native and European beach grasses, the difference of the flat and round stems, the different landforms they create by the way they take root. The native grass thrives in moving sand and will often take hold even down on the beach. The imported grass, like the lupine, was planted to stabilize the landscape, first to protect the railroad to the mill, then for the highway and the houses of mill workers. Settlement here was by no means a sure thing, nor is it taken for granted today. No one is proposing the removal of large tracts of vegetation just because these plants are not from here. It can be costly and painful to be downwind of unvegetated ground. With surprising suddenness, moving sand will fill or undercut foundations, leave roots exposed or bury your garden, get into your house, your clothing, your bed. In the attic of my two-story house, situated on a relatively stable paleo dune a third of a mile inland, I have found fine little drifts of sand.

The imported grass—*Ammophila,* literally "sand loving"—holds the ground all too well. Jack, as we walked in the dunes, remembered when he was a Boy Scout they'd come out here to plant it as a good deed. Probably counted toward a merit badge. But it succeeded so well it choked out native growth and stopped sand from moving inland. It stabilized a naturally semistable environment. This in turn changed the physical configuration of the dunes. Farther inland, along with vehicles, it broke up the original habitat into biotic islands. These islands may be too small for the survival of native plant and animal species. Two plants—beach layia and Menzies' wallflower—have been listed as endangered.

So today's good scouts—known as restorationists—gather to bash lupine and hack *Ammophila,* undoing the mistaken good deeds of the past. And the former scouts, Jack and I, go for long walks in the dunes and the spruce forest as part of an unspoken agreement between us to share healthier pursuits than we have in the past. As part of our personal recovery program, I've practiced these botanical lectures on him because I go to so many meetings and need a place to download all the data they give me. In the past week I've been to three public meetings, two of them on the night of the big wind. For two years I've been a noisy alternate member of the County's Beach and Dunes Advisory Committee, and after four years, I've succumbed to the chairmanship of our local Dunes Organizing Group, which has advised the community on the acquisition and management of its forest and dunes property. And now, besides the BAD committee and the DOG, I've agreed to be a director of our local services district, the public body that actually owns the land.

Of course, now I have no time to just be in the dunes or go to the beach. Spending whole days with a shovel is out of the question—even the paid restoration contractors are talking of fire and bulldozers, the task being so daunting and unending and volunteers becoming scarce. So the only time I'm in the dunes is with Jack, and my few hours at the beach are often urgently convened walks with Joy, in our occasional but so far unsuccessful attempts to put our lives back together. We start out from her place, a few dunes away, or she comes here. My mother is always glad to see her, and the fact that Mom's presence is also partly the cause of our separation gives a bittersweet beginning to our conversations. And Mom stays in the house and puts more wood in the stove and says over and over how she moved here for the wonderful air. "I can breathe up here," she says. "It must be the altitude."

All this is part of the tour I give my guests, the point of the lesson that seems to have come with the territory. Like the gossip that goes with the neighborhood, about the crazy old woman and her strange son the poet. As if I had unwittingly been hired as the damaged caretaker of this damaged land, and in return maybe it would nourish me and somewhat heal the trouble I brought here. We are both subject and object in this matter of "bioregionalism" and have to observe the literal truth that caring for the earth is simply good house-keeping and good hygiene. Just as caring for my mother is taking care of my-self. It stabilizes my life. I'd probably be crazy anyway.

Out on the beach, my guests and I stood before a root wad of redwood, huge and ancient, washed down the Eel River I guessed, higher than our heads. Growing out from between its massive spread of broken-off roots, still holding soil, still upright and somehow intact, a coyote bush—green and alive, a token of the new year, impervious to its doom.

Half a mile south, we cut inland where a dune buggy trail had begun to heal back together. The good news on the beach had been hard to explain—look, no vehicle tracks. You had to see what it was like before, watch a dozen vehicles coming toward you across the hardpack. On the sand sheets of the in-land dunes, the news was still legible—but only a couple of four-wheel drives had been through recently. To the north and south of us, I assured my guests, the signs of damage were abundant and still ongoing. The battle was far from over.

We crossed the dunes to a deep hollow where the willow swamp had suc-ceeded to coniferous maritime forest. We stopped at its edge, looking down the steep dune face. Thousands of tiny grains went sliding from where we stood, down into the shadow of spruce and pine. The dune would bury first the huckleberry and salal, gradually the trunks of trees, slowly killing them. The northern coastal forest is a rare and exquisite place, a small magical world of fungus and fern, bearberry and reindeer lichen, enchanted vernal pools. In the natural course of things, the dune would slowly smother all this delicate beauty. In another hollow, following this slow wave of sand, between us and the beach, another forest is beginning. No one knows what will happen to this process, which we have both hastened with internal combustion machines and slowed with exotic plants. The forest seems unconcerned.

We are in something much larger than us. It is arrogant to think we will "save" what we can hardly even imagine. To illustrate our helpless ignorance of this larger context, I like to throw in a little deep geology on these tours. Of course, I had mentioned tsunamis to my guests, as an obligatory part of our beach walk, and I described the slim odds of getting off the peninsula—or even the beach—before that huge wave arrived with all the force of an enormous submarine upheaval. If you can see it coming, it's way too late. Even more bewildering, if the quake should be underneath us, all the water beneath the sand would be pressed upward. Zillions of tiny particles of sand all suddenly floating independently of one another; the solid land would turn to liquid. To help them visualize liquefaction, I offer the possibly bogus story of a man photographed standing in the dirt street of a coastal Mexican village—he's literally in the street, up to his knees in it. My house, I say, might suddenly be a one-story with a basement. The cat, I imagine, would be the only one able to walk on this jello earth.

I save for last the most spectacular of these coastal "events," as the earth scientists coyly refer to them. As if wind and water had not sufficiently threatened us, a local geology professor and his students found evidence that this entire stretch of coastline—only some three hundred years ago, say 1690s—abruptly changed elevation, wrinkling like a bed quilt into alternating hills and valleys, rising and falling to become ocean bluffs and new beaches. Apparently, it had happened several times previously. Sometimes you went up, sometimes down. Layers of peat were found where there had been a forest, alternating every third of a millennium or so with salt marsh.

I don't have any way to help my visitors grasp the magnitude of such events. The last one was only a moment ago in geologic time, but it is off the scale of imagination. When I dug a new and deeper foundation for my house, I kept coming across a thin layer of charcoal, apparently from the great fire thought to have followed one of these massive subductions. My half acre was not, that time I guess, one of the areas that went under. Of course the OHV riders received this geological news with some pleasure: And you're worried about a few *tire tracks?*

Yes, I say. I want the place to be perfect when it goes.

When we got back to the house, I made some tea for my mother and our guests. It was by now exceedingly hot in the living room, too much for a California January. But despite her obvious success at fire making—a strange exception to her general forgetfulness—Mom kept interrupting to tell me to check the stove, to be sure it had wood. *Jesus, Mom, we're dying in here.* She seemed not to hear. It did little good to ask if she was wearing her hearing aid. Was it

turned on? Was she able to listen? Couldn't she remember from two minutes ago? Of course I couldn't say this. It was impossible to carry on a
conversation.

I was trying to explain to Trina and Sean some of the frustration I was experiencing with what I had once glibly referred to as "reinhabitation." The most
difficult life-form of the bioregion, after one's immediate family, was turning
out to be one's friends and neighbors. Not the rednecks next door, whose differences I'd grown up with and learned to appreciate. No, it was the ones I
thought I most agreed with.

On our way back from the forest, I had pointed out the old elementary
school that occupied part of the hundred acres of dunes purchased by the
Coastal Conservancy. It had also been placed in the custody of the local community and would provide a center for nature study and restoration, a local
gathering place, and a school with the surrounding dunes at the heart of its
curriculum. Here was a chance to empower a small working-class community
that had been run over and left for dead. It would be a place for local culture
and the surrounding land to come together, to begin to restore one another.

Yeah, well—save that for your next bioregional gathering. Some of us who
worked on this project were as naive as those restorationists who are sure they
can fix a slope or save a mountain—hell, save a *species*—only to find the damage much deeper and older and closer to home than they had imagined. It
was a little like my decision to take care of my mother or the certainty with
which my companion and I thought we could deal with her arrival in our
lives. Joy was also for a time a member of the DOG, partly I think because a
meeting seemed the only place without Mom that she could talk to me. Then
it really was the only place we saw each other, and then she dropped off the
committee. There were old issues of control and possession, and the agenda of
the DOG began to reflect personal lives and neurotic interests in power. One
person's visionary project readily became another's job security. There was all
that history. People who had cleaned up the old mill site and made a park of
it—a heroic task, and a watershed event of village history—now felt that this
park should be theirs too. The services district had its own dark history,
though it was only twenty years old—gossip and rivalries, lies, and more or
less blatant threats of violence were still acceptable forms of local decision making. At the meeting last week, I'd sat in the DOG's chair facing twenty people
in boots and black hats who demanded to know where the equestrian access

was in our master plan. We thought we were going to save a hundred acres of sand, but it appeared we had taken on the history of the American west.

I think in a couple of generations, if nothing really bad happens, it will get better. This village is still inventing itself. Democratic community is not something you vote on, but a long-term organic growth. It requires feeding. There is a lot of damage to be lived down. A wave at any time could take us. But I think it has always been fairly crazy—life in the small towns of the bioregion—and probably always will be. Maybe what we're doing is normal.

It often came as a surprise to me when people treated my mother with the reverence of an elder. She had come to seem to me very much like a child, albeit a child going backward in time. My response to her sometimes maddening behavior was not always sane, let alone graceful, and I became certain there is such a thing as contagious senility. Probably the only thing that saved us in this sometimes desperate situation was her still keen literacy and a somewhat fatalistic but sharp sense of humor. She would pick up any piece of writing from the odd range of poetry and prose that passed through the house—Neal Cassady was this month's favorite—and disappear with it into the bathroom for vast periods of time. She was also reassured by my notes to her, telling her where I was or what she was supposed to do. Like, "I'm at work," or, "You don't go anywhere today."

Underlying this sustaining literacy was an older stratum of *dichos,* little aphorisms that would still come to mind for her, and recalled both of us to the obvious: *Todo se acabe en este mundo* (In this world, all things fall down). Or they taught her again (and me too) how to negotiate with the immutable: *Cada día es un mundo* (Every day is an entire world). Or they reminded us that some things, no matter what we wished, weren't going to happen. I would say, Maybe Joy will change her mind: *Y algún día, comerá el gato sandía* (And one of these days, the cat will eat watermelon).

During her long absence in the bathroom, our guests praised me for having taken my mother into my home. This was not unusual, but because this picture of filial piety embarrassed me or didn't really fit the facts, I responded with the story of a crisis we'd had early last fall.

About two in the morning, I was startled awake by a huge *whomp* from downstairs. Joy may still have been staying here sometimes. We were always aware of Mom's being downstairs and constantly expecting the inevitable. When I stayed at Joy's house, it was even worse, with guilt added to the anxiety. So immediately I jumped out of bed and groped down the stairs to Mom's room. When I turned on the light she was lying on the floor beside her bed, flat on her back, in her flannel nightgown and my old blue knit watch cap she wore day and night. She was moaning, "Ay, ay," when I knelt beside her and asked what had happened. She didn't know. We guessed she must have fallen out of bed.

"Are you hurt?" I wanted to know.

"What?"

"Are you hurt?"

She wasn't sure. The smashup years ago, which nearly crippled her, had left her like a glued-together pot. Any re-injury would put a sudden end to her living with me. There were some limits and conditions to our life together, and mobility was a prime one.

But when she spoke again, it was not about her bones. Now it was my turn to say it: "What?"

She said, "I can't *see*. I'm *blind*." I looked at her. Then I couldn't help it—I started laughing. I couldn't stop. Finally, she had to ask me what was so damn funny. It was half a minute before I could tell her. By this time I was lying on the floor too.

"Your hat," I finally got out. "Your hat's down over your eyes."

Lazarus-like, she slowly lifted one arm and raised the cap a little. "Shit," she said. "I thought I was dead." And we both lay on the floor, laughing or crying, crazy or sane, it was hard to tell.

This is the sort of story visitors are expected to listen to, as if it explained something. Sometimes they laughed politely, and sometimes a pained smile was as much as they could manage. It's the kind of story you hear frequently told by care providers, usually to others who deal with damage on a regular basis. Maybe, I guessed, bioregionalism is just rural gothic humor, grown into a movement.

When it was time for Sean and Trina to go, I gave them directions over to the Lost Coast and told them to look for signs of last spring's quake there. But then it's another of those things like my mother's mind, and you had to know what it was like before those ancient rocks came up from the sea. "You all come back," she said, as if she was still living in Texas. Before they left, I got our guests to help me lift the metal roof back on the woodshed. It was actually an old garage door, which I had neglected to attach to the framing. Plenty heavy, I figured. It had blown halfway across the yard, but luckily only some scrap lumber had been in its path. When I told them, "I'll be sure and baling-wire it on now," they laughed, probably thinking I was joking.

Who will ever know, I wondered, how they built the woodsheds, and how deeply they lived in Atlantis?

The Scenic Route

Kartik Shanker

If i had to describe my life or my world at any given point of time, in fact any single given point of time that i can remember, i would probably say,

nuts.
I'm going nuts—which is about normal for the twentieth century—
economically, earning peanuts, or
vocationally, tightening nuts, or
socially, dealing with nuts.

At some such obscure point in time and similarly hazy points in space, i was attempting to become a physicist. Theoretical physics, i am told by the priests of this very metaphysical pornography, is the nearest thing to god. Until it got a little too theoretical. I was a very physical person—that is, after all, why she is, was nuts about me.

but then, i also drive her nuts.
mostly, i drove a bus.

So, i got disillusioned with research. It is not that it was not challenging enough—intellectually and all that. It is not that the marriage with the adviser was a bad one—it was one of convenience and not passionate love (though i did not realize it then), and as is generally the case with the former if not the latter, it had worked pretty well. It is not that i was not up to the task—every priest looked at me as if it were his sperm running in my blood (celibate though they strive to be), colleagues scarred my back with green glowing teeth, others—mere mortals—wept when i condescended to speak with them, conferring on them my quarks of wisdom and cigarette-scented breath. It is not as if i were not in a state of grace. I was, within priestly parameters.

In the end, the choice was not simple. But then, choice, by definition, is not. . . .
Fortunately i had one. I had a bus to catch.

The bus started one sleepy afternoon with my own personal priest, the world-famousinIndia professor leaning too far back on his swivel chair. I held my breath in anticipation of one grand moment, but it was not to be. And he swung back squeaking mildly like a cat with constipation and continued his monologue in C. And i began to wonder why i was there at all, he could have conducted the conversation (and all others) with a cow.

I suddenly notice his forehead, green almost, films of sweat never away for very long, breeding a fungal carpet of worry across his brow—perplexed, puzzled brow—seeking always to unite, as if their union would solve some universal riddle. And i realize that this muddle of bewilderment, ambition, and egotism has wreaked havoc on many an innocent body, wrecked careers, wrenched souls—wretched mind, in search of a grand unified theory of everything—and i realize that i am not, after all, a theory type. Despite god and the word, an atheist.

Shortly afterwards breezelocks kissed me fervently. Under an arch, discussing the purple rose of cairo and astral bodies, i hold her hand, and suddenly she drapes herself around me, windblown hair bobbing, eyes swimming shut, a long and languid kiss, a lasting kiss, a last kiss, alas. That evening and in the years past, the building casts a shadow large enough for me to hide in; the moon singing slowly across a cloudy sky. The moon has been described as the rejected lover; she reminds me not so much of unrequited love as of a deluded academic, a researcher lost in the world of realcloud.
And i ask her why she kisses me.
Because you didn't believe in the grail. Past tense. My bile rises, and in the film of tears, i relive the beginning.

That world is a song from another generation, another incarnation. The notes are still clear; strident, discordant some, surprisingly musical others, church-

bell loud the rest. My future is sparked in one room, memory and emotion crowding together; just another seminar, my fate at the guillotine—my research was then merely an appendage, would become my head.
Roll call:
The priests, bloodthirsty hounds—hunger and habit will incite them to tear another fox to pieces with their contempt— Present.
Colleagues, vulture eager to feed on the remains, with their jealousies—
     Present.
Stormtresses was there too, full-mouthed and fruit-lipped, with her faith intact.

I begin to speak, with boundless enthusiasm, of molecules, sound, atoms,
light, of quarks and strings and condensed matter, of evolution and strategies
and games, of the arrows of time, past and future,
and all the while, the hounds growl and grunt, straining to break line, until
halfway through the talk, one snaps his leash, and the others all break loose
and snap their social nicety chains, and the hunt is on,
they tear through the forest, and the fox hurtles through a blurred landscape;
he is wily, he darts under shrubs, through streams, up trees, he backtracks,
sidesteps, bluffs, but they hound him still.

This goes on until the fox, tired and exhausted, leans back against the tree, as
the hounds come bounding up, baying, with teeth and saliva flashing in the
lamplight. The fox thinks, *Is a fox's life ever different? Is he always dogfood?* Resigns himself to his crucifixion. And then He speaks—the man with eyebrows
instead of a face. He has been watching silently, or sleeping; his reputation is
his armor and missile; his work is monumental, they say; his power is; he mutters four words approximately, two syllables that count, OK.

Without a change in pace, the hounds fall to licking the fox, telling him what
a wonderful chase it was, what a wonderful route he took them through. Full
of self-congratulation, they troop off, patting each other on the back, saying
how they knew all along that fox was really a hound, etc.
And i am egoballoon inflated, not for what his knighthood will do, but that he
gave it. To me. Three years of adulation, prize fox at the fox show three years
in a row; because he liked my one-legged stand on half a half-baked theory?
My puppet act on four disjointed strings? Because i reminded him of the son

he never had? I don't know, but i embark on a journey carrying his bags, in search of the holy theory of everything. . . .

I never did find out (except that now i know it is something akin to bidding for and buying a slave) and so, i remember his forehead, green almost, films of sweat, etc.
I laughed then,
and hardened my throat (my heart having been given leave of absence when i ventured into the crocodile-hands of academia), held swirlcurls one last time, and kissed them all goodbye. . . .
The fox crashes through the forest once more.

Mindmoments later, i was finished with research,
so long and goodbye, mr. fish and chips.

I have always liked buses. Women, it is said, like big, throbbing things between their legs—like horses and motorcycles. Men like big throbbing things wrapped around them. I liked buses. As a child, i always sat next to the driver, getting sublime pleasure from his terrifying power. He was a god, with all that power invested in his anointed feet.

So i became a bus driver. Mintbreath would later classify me somewhere between a nematode and an annelid worm,
better a free-living worm than a parasitic one.

I drove these magnificent creatures all over the countryside, and my heart throbbed with them. I can still feel the pedal beneath my foot. We enjoyed many a fugue, the lights leering lecherously at trees, crashing down the highways, exploding through innocent hamlets, reaching exhausted, enervated some faceless, nameless end. *Highway throbbery, all the way.*

She was way behind in a maruti zen, red, bobbing up and down in a sea of evening traffic like a little sailboat, paper perhaps. I maneuvered my dinosaur through impossible roads, but in the rearview mirror, caught tantalizing glimpses of her; asphyxiating shirt, strained-back hair, gray pants or skirt, i guessed. And i decided she needed saving, from herself; watched relentlessly, keeping her just behind me, obstructing traffic and making her blow her horn and top.

At the next junction, she was out in a flash, treating me to a barrage of fluent invective and the vision of a beautifully angry woman; my passengers, a motley crew of the old and very old, of heart and mind if not body, tittered and blushed.

Sorry ma'am, you'll have to speak up, i tell her in my best oxfordspeak (this in a country where english—like cheese, cars, and chocolate—is the privilege of the very privileged).
She looks at me, her eyes popping, her voice, once a babbling brook, reduced to an autumnal gurgle way back in her throat.
—Could you say that again? she ventures.
—You're stopping traffic, i reply.

She blushes, so i jump down,
Let me not, to the marriage of true minds,
put my arms around her,
admit impediment.
and kiss her. Softly. Gently. And forever.
Goodbye sweet lady, goodbye

I jump back up to a barrage of horns and the tchtching of passengers, and drive away as the lights turn red, leaving behind a dinky car with a dinky person who is a minty breath of air in my diesel-fumed life.

And i tire of ferrying mental, emotional corpses from one city to another; all i see are faustian salesmen, permanent adolescents, incompatible couples. Too often, i see their scars, as they sit by my side and talk into the night, over the drone of the engine, over the snore of my belly-up conductor, while the cleanerboy looks wide-eyed into the night and the road ahead. I never did hear a happy story there; of course, the successes and the dead lie smug and dead in their seats. Only the guilt-ridden, the shame-spent, the rage-driven rise, like cream or bile, to the surface of my four-wheeled centrifuge.

And then, occasionally, i show off (isn't that the whole point?), the bus driver shocking with his shakespeare, charming with his chaucer, pardon my french, etc.

With a bunch of pretty girls,
—What on earth are you doing driving a bus? (giggle giggle, etc.)
Ever tried it? Unbelievable orgasms (now this is almost true).
Twitter giggle again.

Intoxicated by snowy breath, i swap the public service for a private one; i ferry schoolchildren to faraway places, *let's take the scenic route*
to places with history rather than industry,
beauty rather than cruelty,
richness, of earth and water,
wide open spaces,
temples, mosques, and churches,
forests, fields, beaches, and backwaters,
and in one of those places, the seed is planted.

The kids i meet are mostly kids, as they tend to be, ghastly and angelic in ir-repressible combination.
—Excuse me, can you put my bag on top, please.
—Sure thing, heave ho and off you go.
Buzz buzz buzz . . . He speaks english . . . Chatter chatter . . . He's a physicist
. . . He listens to . . .
—Of course i listen to jethro tull, *and jimi hendrix and jim morrison and mozart, beethoven and bade ghulam ali khan, i know you haven't heard of them, before your time, but if you haven't heard the grateful dead, sergeant pepper, and pink floyd, i'm going to have to educate you.*

But i got educated, although i didn't realize it till later.
They loved me, of course, and i love them.

The teachers eye me with wonder and distrust; *is he an angel or just the devil in drag? ohoh hoity-toity PhD from youknowwhere; then why is he driving a busload of kids from smokestack to nowhere?*

One day, i see a familiar red bobbing object. I drive alongside, and we stop simultaneously at the inevitable red. Red for passion, safety, irony. I hop out and run across; precocious mindreaders, my children sing
he loves you yea yea yea . . . and then
bright are the stars that shine . . .
The beatles education has been worthwhile.

I ask her to dinner, the lights turn green.
Two weeks later, i move into her house.

Life is good; we live together; coolbreath, a hyena, and i. The hyena is waiting for the lion (i have long since graduated from fox) to have his fill of the gazelle. Knowing that lions, being lions, will get lazy or bored and leave the feast midway. The hyena will then feed, for once the heart is gone, the rest is easy meat. We live in a very strange jungle.

I try to understand it.
—So why do we all still try and survive in the smokestack?
—Civilization, culture, high living, he cackles.
Smoke, luxury, cheap thrills.
—Education, art, good food, she adds.
Brainwashing, carcinogens.

The seed has indeed been planted; cotyledons swell into neocortex, plumule snakes down medulla, and radical roots into concrete cranium.

I do not notice. Yet.
I drive my bus.
But then, buses are not forever. If you've lived out one bus, you've lived them all. I seek change. If you're going to marry a way of life, you need variety. I want variety in my co-throbbers—a throbbifarious life. The hyena suggests that i become a car mechanic. I can test drive all the cars that i fix.

Then i became a car mechanic
in one of those very uppity localities.
Where folks in stuffed suits and strangler jeans, bring in their mercedes (what
is the plural), toyotas, porsches, fords, hondas, ad nauseam. Where the people
who bring in their etcetera are too ashamed to admit that they have a maruti
suzuki at home.
I would lie under these delightful, different, delightfully different animals, lis-
ten to them throb, tinkle, and tinker, . . .
while they told me their troubles with their cars, their jobs, their wives, kids,
dogs, parrots, *when sorrows come, they come not single spies but in battalions*. . . .
To blank faces, i'd say . . . hamlet.
To blank faces, i'd add . . . shakespeare.
To blank faces, i'd finish . . . nevermind.
It's a common enough line, i know some shakespeare, that's why i am a mechanic.
Winterbreath says, have some sense.
No beast so dumb, i tell her, but has some touch of sense,
but i am no beast and therefore have none. Dick the 3rd.

She ignores me, but i don't blame her. She's this up-and-coming executive at
this large, slick bank. She goes out every morning looking like an air hostess,
ruins the lives of several thousand people, and comes home wearing a mother
theresa face on a banana-peel body. What's a worm doing in her life, even if it
is a free-living one?

Until.

Today, this car glides kestrel-clear into my garage, and hovers on its cliché—
gossamer wings over me. Today, it touches me and

it is a saab. You don't see many of those here. A woman steps out. You don't
see many of those either. A woman with exquisite breasts. Some women have
exquisite legs, others have exquisite faces, hips, bums, necks, backs, stomachs,
. . . minds. This one had exquisite breasts. It's all hormonal; beauty, after all, is
in the loin of the beholder.

I tinkered, then rumbled with the saab. I test-drove the saab. Then i test-drove
her.

I told arcticbreath i was leaving. Surprisingly, she was sad. Though she had always expected it, she said. She wreaked devastation in the lives of monkeys, rats, dogs, insects, crabs each daylight hour, but her soft spot was worms. And i, a part free-living worm, was going to become a parasitic one. The worm turns.
The turned worm. The toasted worm. The tasted, tested, tried worm.

We—the me'm saab, the murmur saab, and i—drove away into the sunset. We got burned, but then it got dark, and we didn't see the heat anymore.
And i tinkered, heaved, and throbbed
with buoyantbreasts and the saab.

I become her toy, the curio she shows off along with
her terracotta horses—from northvillage, you know—
her earthen pots—from southtown, you know—
her paper lanterns—from china? i don't know—
her tropical fish—from the river, you'd never have guessed—
her bulbheaded cane from darkcontinent, my phallic symbol.

my appendage, to be waved and flashed.
my silver cap, which glints in the lamplight when she smiles.
my diamond ring, which catches chandelier rays when she gestures.
my kashmiri shawl, to wrap around her.
my persian carpet, to admire, worship, walk on.

Don't choke on that, i tell her spaniel one evening, who cockers an eyebrow at me questioningly, chews his bone, will nuzzle me affectionately later. And i discover it is the most honest, loving, caring conversation i have had in days. Another discovery that will lead to another . . .

Meanwhile i get increasingly intoxicated, charming her charmed circle, of diamond necklaces on wrinkled skins, pipe-smoking glazed eyes, bandy-legged ex-jockeys, of tranquil women, comatose with servitude, who expect me in their boudoirs.

Oh, you have a bmw, bring it in sometime, wear a shortenough skirt and i'll service
both of you.
And i drink it all in. Until society is oozing helplessly out of my nose and ears.

To mr. poshpants and mrs. bigbust, she says,
you must meet him, he used to drive a bus, playing sylvia plath to my ted hughes,
such a passionate romance, such a nice fantasy;
and for this performance, i wait grime-covered, squatting in the balcony, wrap
my lungi, flick the beedi away, and waltz into the room.

Sorry sylvia, i didn't know we were expecting guests today

And then return, in tight jeans, crotch and mind bulging simultaneously, quot-
ing nietzsche and kierkegaard,
caan and kant, and kuhn.

Later, grumbling gently like an idling engine, she'd turn her mind to nuts and
bolts,
grime and grease, filthy hands playing the piano,
paan-clogged mouth singing pavarotti,
oily hands rubbing two wires together, the engine roaring under their caress,
and have her orgasm.

And all this time the plant grows,
sending roots into my arteries, rootlets into my capillaries. Flowers bud and blossom
and fall unnoticed. Branches turn inward and crisscross each other, form a network
of infinite complexity, a million synapses buzzing with energy. And yet a million other
tips strain against concrete, osteocement, convention, cranium,
until finally, exploding
fragments like glass bursting, like missiles,
in one nuclear release
the dryad stretches his arms upwards, outwards, relishes his freedom. And so i told
regalbreasts i was leaving. She smiled enigmatically—she was, is the only
woman i know who can smile from her breasts—and said,
I have a surprise for you. You conceived, she said.
I was shocked.

She took me gently by my eyes, held firmly in her bosom, and led me to the door.
There in the maternity garage, glowing in the early morning sun, or late evening sun, or early night sun, or late night sun, or something moon,
stood a baby saab. Breathtakingly something or the other.
A bastard baby. My child.
The proof of my potence.

My child and i, we went away, trembling together, to the wildnesses, far from the madding, stifling, virulent, viperous, vicious side of the coin that had bred, nurtured, maimed, and nearly murdered me. We went to the virgin, pristine side; it isn't even a side anymore, just a route. But we drove round and round it again and again. We heard the deer honk, the trees laugh, birds giggle, nature guffaw, fires cackle, untroubled, undisturbed, each living out its natural life. And we slipped into it unnoticed. We saw—actually saw, not just knew that it had to be—the sun rise and set. We saw the rivers flow, felt them, heard, smelled, and tasted them flowing, unchecked and unrestrained; forgot their cousins lying wheezing, choked and strangled in our past. We knew the stars at night, sensed them during the day, and learned the language of the universe. This one at least. There are others, i believe.

We filled the spaces with our songs,
flung freedom's frisbee with abandon,
flowered in the spring,
bloomed in the morning,
floated in autumn,
spread our petals at dawn, and let the dew settle on our wingtips,
permeate us, till we dissolved and became one with the streams that tickled
the earth till she giggled, the rivulets that joined hands and playfully stroked
the earth,
the swelling, tripping, youthful whitewaters that dug into her sides so deep,
until she pushed them over with a guffaw,
the motherrivers that gently and expertly played over her breasts, her bushes,
her gorges,

caressed her cheeks, her crevices, and kissed her navel,
mighty riverseas that flowed over her, enveloping her in their spirit,
until she released them—and herself—in one maternal sigh.

We sat by the rivers the most; they did not believe in gods or grails, only created them, mirrors of the universe, the grand canyon, the amazon delta,
ganga's banks.
We watched them rise and fall with the monsoon, heard their secrets at night
whispered on the wind, but like two molecules of a butterfly's pheromone, it
was enough to draw us to them.
We worked our way along the rivers, from creation to nirvana, along a history,
through an evolution, across a geology. Starting at the trickling lines, wet rocks
on lonely mountaintops, we followed the story through its winding tale, meeting friends along the way, each of their stories an epic, beautiful and whole.
Every water drop thanked us,
and we thanked them, the little saab and i.
I knew my child like i know myself, perhaps better.
As with most parents, i was even more sensitive to the moods, ailments, pleasures of my toilchild than my own.

The dryad has set root
firmly now. Her roots are but my own, her branches and stems are but my neurons, i
am dressed in her foliage.
And so i begin to shed skins.
Great masses of loose keratinous folds slough off; the first one wrenches, as though it
were wound tight around something like a soul; something moldy, bloody, caked with
dead cells, clotting with past drains out, a menstruation and a molting, a cleansing
and a catharsis.

The first one is the hardest (or so they say), each layer comes off, drains, sloughed
with effort; the dryad breathes, and, like the circusman-breaking-the-chain-around-his-
chest, something gives with each breath.

And all this time, my child is with me, in spirit, all the way. The silentchild
doesn't say much, but takes everything in, feels it with great sensitivity, feels
my guidance and passively accepts it. For the moment.

Is that where it goes wrong?
Are we wrong, after all, to guide? To even try?
Perhaps we are wrong to guide even ourselves,
we should flow like rivers.

My sweatchild was born into a different world.

So just as i was finally being revived and resuscitated by this unbearably truthful beauty and this intolerably beautiful truth, i felt a certain uncertainty, an inevitable note of discord. My bloodchild was not happy. The damnedchild was
used to the ways of anotherworld. Our breathing was no longer in synchrony;
joints creaking, lungs heaving, feet worn out, a heart aching for the other-
world. Try as i did with patience and tender care, my soulchild would not listen and eventually rebelled. As i sat, resigned, tired, and sad, shedding tears
with the moon, i knew it was but a matter of time. So i let the last skin slough
off, the giftwrapping of my soul, and gave mychild freedom. Left open and
free to leave.

The next morning the worldchild was gone. Back to a world peopled by others
of otherkind. Leaving me, as i had left so many others.

That of course leaves me and the river.
A world peopled by smells, molecules, sounds, atoms, light, quarks,
the present.

But ultimately that isn't enough for me.
I may have the quintessential quark.
I also have a quirk.
Unlike siddhartha, i need a significant other to finish this story. Technically of
course, a man, a lobster, a rock, or even a monitor lizard would do, but on
the whole, i'd prefer a woman. *There is something about the idea, brought up as
we are on dualities and dichotomies, there are some things that are too deep to shed.*

And today, she has come.

I see her materialize on the other bank, ethereal, ephemeral, eternal. What
strikes me first is that she is wearing nothing—no clothes, attitudes, beliefs, ex-

pressions—not a thing of the surface that i can see. It doesn't merely strike me—it knocks me back through two bushes and a tree. I recover and move forward, step into the river—naked, for i too have shed my trappings a long time ago. As she steps in, she looks at me and smiles, serenely. I stop, for a thoughtmoment. I think it is breezelocks, mintbreath, buoyantbreasts, but then i realize she is all the women i have known and not known.

We move forward, gently together, in absolute harmony. The river strums gently around us, the air exhale-whispers a tender sonata, stars fall, fade, the universe orchestrates a symphony that rises to a crescendo, as we meet halfway

and flow and *flow* and flow.

Parlor Game

John Farris

I don't know what it was, but it flew—requiring some two inches of space as it drifted lazily through the cabin—exploring, inspecting, perhaps looking for a mate. It looked like one of those male mosquitoes, only it was larger and seemed to sit comfortably in the air like a pilot rather than an insect. Alarmed, I asked my friend if I should kill it, and when he shrugged unconcernedly, ambling off to his room like the insect, he being tall and thin, with legs like rapiers and bent at the knees, I felt ashamed of myself for being a city boy and did not kill it, though I could not help being glad I was not its prey or mate, as streamlined as it appeared to be and built for war. It was a lazy afternoon— having just turned summer—and the slope leaning down to the valley this side of Mount Wildcat was blanketed with clover. Bees like little milkmen in stripes made their collections and deliveries of sap and pollen between the daisies and clover while gaudy butterflies fluttered by like vain suitors, preening themselves on every blade of grass before taking off again, unconcernedly. A woodpecker, deciding a metal communications pole might yield a mealy grub or two, slapped a resounding tattoo (not unlike the tumba the Yoruba use, or is it the Ashanti—I don't know, it's been so long—maybe it's the Fang) that was amplified by the hollow metal so it reverberated through the valley— tumba tumba tumba, rat-a-tat-tat—before being joined by the trilling sopraninos of warblers and the piccolos of the bobwhites and ultimately by the thumping bass of a large bullfrog from the vicinity of the pond and the incessant, maracalike buzzing of other less attractive, less welcome winged

Bob Braine

specimens exercising their right to make music too. Earlier that day I had seen the most beautiful spiderweb, its delicately laced patterns of hexagons defying logic, hung like a suspension bridge outside the cabin window from the middle sash to the sill. I was taking note of how mathematically precise it was when a ruby-throated hummingbird suddenly thrust itself into the view-finder of the window like a helicopter with a forward gun before darting away just as quickly. I wished it had come to see me and that it had stayed a little longer, but it made only that appearance and was off. I never saw it again.

My friend said his family owned the land because it was worthless—being all slope as it was, all seven thousand acres of it—and covered with hemlock. "Hemlock," he said, "is worth about fifty cents if you don't count the aesthetic value or its intrinsic relationship to the biotic community," he being an ecology junky teaching fly-fishing at the YMCA in Frost Valley, an intricate business involving just the right lure, a proper cast, and a thorough knowledge of how the animal will rise; "kiss rise" or "tail rise," it being a predator whose father had invented the gas mask. I had to laugh in appreciation of his qualification

Bob Braine

that no land is worthless unless someone says it is, and that might be because
they want it for themselves, like Saudi Arabia or Qatar or America, because
you had to appreciate the vastness and solitude, and that the earth was yours
from hell-deep to heaven-high.

There weren't only hemlocks. There were silver and white birches, there
were ash, larch, and pine and even a little cherry tree to the left of the cabin,
almost out of sight. (So what if it didn't bloom and bear fruit, it was so very
pretty, and at the very least, the wood was good.) There was a pond too, all
the way down the slope, almost to the base of the mountain with a stand of
hemlock behind as black as any forest in Bavaria. There were bullfrogs. Pond
skimmers skated across the surface at their peril. The pond was opaque, fed
from the runoff of winter snow, the water en route to the reservoir nine miles
to the south just outside the town of Neversink, having to slink below Claryville (for, as my friend said, there hadn't been any blacks in Claryville for as
long as he could remember, but now there was one family and a few American

Indians—they incidently having never occupied this valley before, it being too cold, too mysterious—so much so that there were many superstitions and many taboos regarding it). Deer were so abundant that myrtle had to be planted around the edges of the garden to keep them away from where my friend had planted in the center, more for their own good than his, opium poppies and marigolds. The large blackberry patch on the south perimeter hid rabbits and birds, the foxes that came for them, and snakes that took the eggs of birds foolish enough to lay them there, like the woodcock and the inland plover. The little shack, which used to be the outhouse when the cabin required one but was now the guest house for hardy and insistent visitors willing to pay for the privilege, leaned precariously to the right of the pond, while just behind and to the left was another stand of ubiquitous hemlock that my friend said he was going to clear so as to get more sunlight for the garden in the late afternoon. As it was, the cabin itself got plenty of early light, being on the eastern slope of the rise and facing that direction so it was possible to watch the sun begin its antipodal journey to its false dawn. By nine, the light would sharply outline the many shades of green of the grasses in their lines of demarcation up the slope and throw the hemlock behind the pond into silhouette, rendering them dark—dark as the myths that had once surrounded them. These mornings would find me sitting on the little terrace on my side of the cabin, taking in this phenomenon with my painter's eye, how where a thing was gave it its color, and how without the sun everything was black, black; drawing: the bleached and rumpled cumuli like gobs of wash that had decided they would shake themselves into the loose shapes of animals; bears and bison, oryxes, lynxes, palominos, great eagles, and such. If it rained, I would throw open the glass door leading to the terrace from my room, place myself in a big easy chair that my host—the trout fisherman who had the appearance of a fly (being ever solicitous of my comfort) had provided, and watch the mist rising from the silhouettes of everything, drawing, feeling I had finally joined the Hudson River School, though it wasn't the Hudson I was looking at but Wildcat Mountain, to which the lynxes that hunted at night in the blackberry patch had given their name.

Night belonged to the fireflies—millions of them like tiny low-flying stars nodding and twinkling everywhere, forming thousands of constellations, mist framing these minuscule galaxies as they formed and regrouped, formed and regrouped in their yellow-green phosphorescence, mirroring the inky sky with its millions of twinkling and shooting stars like tracers in a celestial war, zinging to their appointments with destruction, blazing their way in the briefest moments of glory, or like being at the inky bottom of a sea channel, two miles down; the phosphorescent creatures swimming and blinking like characters in an H. P. Lovecraft story where overstimulation of the pineal gland allowed one to see them streaking through the atmosphere, and the moths hovering at the cabin window where I usually sat, hankering after the artificial light of the Sylvania light bulb in the lamp just over my head, and a hundred thousand midges: on the night of the day in question, the day I had seen the strange insect that looked like a pilot (and who uncannily resembled my host), I had been sitting and watching this little macabre dance of the moths and other insects—in particular, some little green ones with articulating thoraxes and what looked like the tiny things' antennae that they constantly threw from side to side like little whips—drawing, when the jazz I was listening to—Dexter Gordon, Johnny Griffin, and Sun Ra that my friend Alan (who, come to think of it, also resembled a giant insect) had given me for the trip suddenly stopped, the tape having finished. Bending down to reverse it, what did I see almost as if on cue, locked in a fierce struggle with a spider web, but the insect—the one that looked like a pilot and built for war. Only now it had been captured by the silken strands of the web from which it had managed to free all but one slender leg, which was stretched taut, straight out, the medium in a deadly tug of war. It was turning and twisting like a mad dancer and was going to lose at least a leg at any moment. Thinking fast, I freed it with my cane and after shaking itself as if in disbelief, it slunk off, perhaps to recover.

The next morning, while planting myrtle around the garden, I saw—in the compost heap next to the blackberry patch—a small stump with roots that appeared to form the insect's face. It was of a species resembling mahogany unknown to my host, who now walked with an odd gait, like a limp. I took it in. It has since formed the head of one of my greatest sculptures, with feathered oak planks for wings, a brass abdomen and thorax, and six of the thinnest wrought-iron legs imaginable.

The next day I left for the city, my friend driving me the half hour to the bus stop in Liberty, thanking me profusely for my visit: I keep the insect in my house, where it reminds me from time to time of my fear.

Jaanika Peerna

Epilogue: Places in the Wild

The trick is to tell these stories so you're not the center, so the place is in the center—not the place you come from, but the place you will go. The wild, the center of your unease, urges you to move on.

Four moments, four places in the world:

I was driving with a millionaire over the Splügenpass. That's up from Switzerland, down into Italy. On the rise, a windswept mountain emptiness, strangely silent, all planned out, a parkway over the edge of order. Descending, hairpin curves to the valleys of celebrated lakes. Plaster churches, stone towns.

The journey took many hours. Occasionally we would stop in bars in the middle of nowhere, buy everyone a round, and then leave. "It doesn't matter where this money comes from," he said. "It's easy to see where it goes."

Across the Thorung-la in Nepal, the trail weaves slowly upward, six thousand feet from the churning river below. I met a man who said he'd carry my load as far as I wanted to go.

"You don't work for me," I said. "Let's carry our loads together."

The only way of crossing this pass was to walk, just as the goods have made their way back and forth, up and down, for hundreds of years. I asked him if he ever tired of the trip, and he replied: "It's not as if I'll end up anywhere. I cross these mountains to see how far I need to go before I forget the look of my home."

In downtown Atlanta no one walks the streets at night, because all the shiny new buildings are linked by tunnels way above the ground, which insulate people from the sky. On the empty sidewalks below, rendered invisible in the fog, I met a wanderer who had nowhere to go.

"What's it like on these evaded streets?" I wanted to know. "You're the only other one out here."

"The good people live on the edge," he said. "There's no one worth knowing in the center."

The island of Troms is the exact size of Manhattan, and on it is the world's biggest city above the Arctic Circle. I sat on a mountain just to the north of it, gazing down on the whole snowy island spread out like a map below.

A local told me that the view would not look like this tomorrow. Crowds were streaming in. There was no more room. Seventy thousand people on the island.

"Is that a lot or a little?" he asked.

I talked to him of skyscrapers, apartment towers lost in the smog. Cars at a standstill in all directions twice a day. He saw the future, I saw the past.

I spent all I had crossing the peaks of the world, for reasons hard to explain. It was never hard to make people smile, though. They kept asking: *why did you have to come so far?*

I looked down.

There were always more people, and always less land. And the wildness crept into us. I saw it in the eyes of everyone who climbed up to consider another side. *Any* other side, any place different from where they were. After they climbed up, after they looked over, after they looked back, this is what they said:

> *You have to stop replacing senses with words.*
> *You must look around to see the shapes, the splits,*
> *the shadows that happen because no one planned them*
> *and which no one will ever completely know.*

In the wild there's always something to run from, things to scare you, dangers not in the dark but in the feverish light, the inescapable brightness that will not leave you, that will not revert to any quiet or peace. Everywhere people are descending and ascending.

We're on that course. And we look back in peril. We see a nature that draws us toward it, even as we can't understand it. Were we to understand it all, we would have made it our own. And it would no longer be more than us, but circumscribed within the lines we have drawn. It's safer to be in a world that knows more than we do. That's a world that can be trusted.

The crossing, the climb, the walk, the drive. All over the world there are things we cannot see from where we are. We can move toward them but they'll still be just beyond our grasp.

People all over the world roaming. This one laden down crossing the Himalaya at dawn. That one in the city center where no one can live. And the millionaire driving through the Alps. All are looking for those places we'll never reach.

Still, there are things that make our hearts race. There is a love of wildness that will want us to lose control. It's on a highway median, down a dead-end street, or on paths that lead nowhere all over the world. We can find it even if we don't leave the house. From any spot there are passes to other worlds, trails that start from our front door but won't let us back in. We can choose our route. But the destinations are all the same.

At first, when you look at this new world, it seems just like the old. There are rivers, towns, roads, and homes. But then you look again, and you realize you've been fooled by the mist. You see that the rivers are dry. The towns are empty. The roads are falling apart. There is a dryness in this new place because it hasn't yet been tried out, lived with.

There is a desert on the far side that hasn't been crossed. This wildness is no place to be stranded. You grow thirsty just wondering how long you'll survive there before you give in, when your liquid memory will have been sucked dry and there is no sustenance left in the things you thought you knew how to use. You'll have to make unfamiliar foods out of the sand and create a life out of a whole new, empty ecology. It will be nothing like your life today, or yesterday, or the day before . . .

But you can look back. Don't be afraid. Then look ahead. Stop. Relax.

The new land, the new range of ways to be human in the world, the new levels of understanding we can bring to and tap from nature—that is the one that will endure.

—David Rothenberg, *editor*

Contributors

Rick Bass lives on a remote ranch in Yaak, Montana, and swears that he did in fact see plenty of bears in Romania (though he never did get to smuggle any out). He is the author of numerous books including *Oil Notes, In The Loyal Mountains, The Ninemile Wolves, The Lost Grizzlies,* and *Where the Sea Used to Be.*

Charles Bowden started writing because what he was reading didn't cover what he was seeing and hearing and feeling. He had no schooling for the work, just the appetite. Bowden has worked on a newspaper, freelanced, started up and run a magazine, and scribbled a dozen or so books, including *Frog Mountain Blues, Desierto, Trust Me: Charles Keating and the Missing Billions,* and *Blood Orchid,* from which we excerpted in our first issue. He writes what is called nonfiction: "Everything I like is true whether it happened or not."

Bob Braine is an installation artist and photographer whose projects investigate wildlife and ecosystems in urban environments, most recently those of New York City's East River and the Elbe River in Hamburg, Germany. His work is shown in galleries and museums in the U.S. and Europe.

Adam David Clayman is a documentary photographer residing in New York City. His photographs included in *Terra Nova* are part of a long-term project documenting sacred sites in India.

Doug DuBois's photographs have appeared in *Artforum*, *Geo*, and *DoubleTake*. He has received fellowships from the National Endowment for the Arts and the San Francisco Foundation.

John Farris is the editor of *Digitas*, a digital review of arts and literature. His poetry, fiction, and jazz criticism have appeared in numerous anthologies and art catalogues, and a collection of his poetry, *Not About Time* (1993) was published by Fly By Night Press.

Mariana Kawall Leal Ferreira is a medical anthropologist who conducts research on health-related issues in Southern Brazil and in Northern California. From 1978 to 1984, she taught school on Brazilian Indian reservations. Since then, she has been a consultant for Indian organizations in the Amazon. She has published several books in Brazil.

Lynda Frese is an associate professor of art at the University of Southwestern Louisiana. Her widely exhibited photo-assemblages and computer collages examine the dynamics of memory and pose questions about our changing relationship and experience with place and the sacred through time.

Ray Isle was born and raised in Texas but now lives in New York City, where he works for a wine importer. His stories and essays have been published in *Ploughshares, Agni, Fish Stories,* and other literary quarterlies, and he is currently at work on a novel.

Photographer **Algimantas Kezys** was born in Lithuania and has lived in the United States since 1950. Several books of his work have been published, and his photographs have appeared in magazines on both sides of the Atlantic. The photograph in this volume is from his collection *Būties Fragmentai* [Fragments of Being], which will soon be available in English from Galerija.

Jaron Lanier is a computer scientist, composer, visual artist, and author. He coined the term "virtual reality" and was a principal pioneer in the scientific, engineering, and commercial aspects of the field. As a musician, Lanier has been active in the world of new classical music since the late seventies. He is a pianist and a specialist in unusual musical instruments, especially the wind and string instruments of Asia.

C. T. Lawrence's fiction, essays, and poetry have appeared in numerous journals and anthologies, among them *The Best of Writers at Work*, *Western Humani-*

ties Review, *Ekphrasis*, and *Puerto del Sol*. She teaches writing, and is finishing a Ph.D. in fiction at the University of Houston, where she was awarded a Donald Barthelme Memorial Fellowship in 1995. This year she received the New England Writers Award for short-short fiction.

Jerry Martien is the author of several chapbooks of poetry and *Shell Game: A True Account of Beads and Money in North America* (Mercury House, 1996). He is completing a book about the caretaking of people and place—particularly, his aged mother and the coastal village of Chicken Beach.

Raymond Meeks's photographs have appeared in *Outside, Rolling Stone, GQ*, and *Men's Journal* and have graced numerous book covers and CDs.

Errol Morris is one of our most talented and original filmmakers, most known for his 1988 feature documentary *The Thin Blue Line*, probably the only movie to have overturned a murder conviction. Other films include *A Brief History of Time* (1992), which explored the work of physicist Stephen Hawking, *Fast, Cheap & Out of Control* (1997), and his forthcoming film *Mr. Death: The Rise and Fall of Fred A. Leuchter, Jr.*

Ethnobotanist **Gary Nabhan** is the author of *The Desert Smells Like Rain* (North Point Press, 1982), *Songbirds, Truffles, and Wolves* (Pantheon, 1993), and *Cultures of Habitat: On Nature, Culture, and Story* (Counterpoint, 1997). He is also the Director of Conservation and Science at Arizona-Sonora Desert Museum and a MacArthur Fellow.

The late **Bikram Narayan Nanda** was professor of sociology at Jamia Millia Islamia, New Delhi. He passed away in an accident in 1994. (*Terra Nova* expresses deep regret at his tragic and premature death. This was the first essay we accepted for the journal, in the spring of 1994, when Bikram was a Fulbright fellow at the New School for Social Research in New York.)

John P. O'Grady served as *Terra Nova*'s poetry editor for two years. He is assistant professor of English at Boise State University and the author of *Pilgrims to the Wild* and *Grave Goods*.

Jaanika Peerna is an artist and educator from Estonia, now living in Cold Spring, New York.

Ted Perry is the Fletcher Professor of the Arts at Middlebury College in Vermont, where he teaches courses in film and video. Most recently, he edited and wrote an introduction for *Antonioni: Poet of Images*. His new book, *Movies, Me, and Us*, will appear this fall.

Val Plumwood survived the crocodile attack in February 1985. She has recently held visiting professorships at North Carolina State University and the University of Montana, and is currently an Australian Research Fellow at the University of Sydney, Australia. Her most recent book is *Feminism and the Mastery of Nature* (Routledge, 1993), and she is working on a book entitled *The Eye of the Crocodile*.

D. L. Pughe is a writer and artist living in Berkeley, California. Her essay "Letter from the Far Territories" was published in *When Pain Strikes* (University of Minnesota Press, 1999), and two other passages from *A Philosophy of Clean* can be found in *Nest* magazine (1999).

Kartik Shanker thinks he's becoming an ecologist, having worked in remote forests on rodent communities and walked windy beaches in search of sea turtles. At heart, though, he's probably just a writer. He currently lives in India.

Mohammad Talib is a reader in sociology at Jamia Millia Islamia.

Jerry Uelsmann's work can be viewed in the Metropolitan Museum in New York, the Fogg Art Museum at Harvard University, and the Royal Photographic Society in London, as well as other major collections. He has been graduate research professor of art at the University of Florida since 1974.

The Future of *Terra Nova*

The publication of *The New Earth Reader* marks the beginning of *Terra Nova*'s new life as a book series. We are presently expanding two of our previous special issues into books. *The Music and Nature Book* will be published by Wesleyan University Press, together with two audio CDs. It will be the first anthology ever published on this fascinating subject, based on our summer 1997 issue. The University of Arizona Press will be publishing *The World and the Wild,* demonstrating how wilderness is being preserved all over the planet, especially in developing nations, an expansion of our summer 1998 issue.

Next year will come our first completely new volume: a fat, juicy collection on the theme of *water:* steam, vapor, rushing, frozen, necessary, cool—with our usual insightful and unique range of contributors. After that, we hope to publish one new book a year. Earth? Air? Fire? Let us know where you think we should go.

—David Rothenberg and Marta Ulvaeus, *editors*
terranova@njit.edu

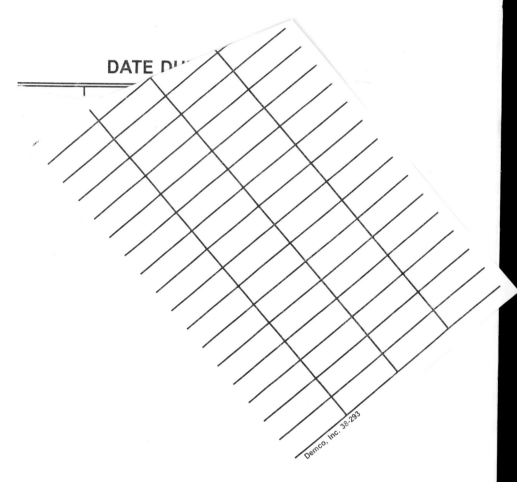

DATE DUE

Demco, Inc. 38-293